JavaFX
应用开发教程
——基于JDK 9与NetBeans实现

宋波 ◎ 编著

清华大学出版社
北　京

内容简介

本书介绍JavaFX的GUI程序设计的基本内容,精心挑选并深入探讨JavaFX中具有代表性的应用开发技术——JavaFX Media、JavaFX 3D、JavaFX Web、JavaFX-Swing、JavaFX 图表,以及JavaFX 动画与视觉效果等。书中每章都有大量从简单到综合的示例,对重点示例阐述编程思想并归纳必要的结论和概念。本书的电子教案及源代码等配套资源均可在清华大学出版社官方网站免费下载。

本书可作为高等学校计算机、通信及自动化等专业的教材,也可作为相关专业技术人员的自学参考书。

本书封面贴有清华大学出版社防伪标签,无标签者不得销售。
版权所有,侵权必究。举报:010-62782989,beiqinquan@tup.tsinghua.edu.cn。

图书在版编目(CIP)数据

JavaFX 应用开发教程:基于JDK 9 与NetBeans实现/宋波编著. —北京:清华大学出版社,2022.10

ISBN 978-7-302-61499-9

Ⅰ. ①J… Ⅱ. ①宋… Ⅲ. ①JAVA 语言-程序设计 Ⅳ. ①TP312.8

中国版本图书馆CIP数据核字(2022)第139344 号

责任编辑:郭　赛
封面设计:杨玉兰
责任校对:焦丽丽
责任印制:沈　露

出版发行:清华大学出版社
　　　　网　　址:http://www.tup.com.cn, http://www.wqbook.com
　　　　地　　址:北京清华大学学研大厦A 座　　邮　　编:100084
　　　　社　总　机:010-83470000　　邮　　购:010-62786544
　　　　投稿与读者服务:010-62776969, c-service@tup.tsinghua.edu.cn
　　　　质量反馈:010-62772015, zhiliang@tup.tsinghua.edu.cn
印　装　者:大厂回族自治县彩虹印刷有限公司
经　　　销:全国新华书店
开　　　本:170mm×230mm　　印　张:17　　字　数:314 千字
版　　　次:2022 年11 月第1 版　　印　次:2022 年11 月第1 次印刷
定　　　价:59.90 元

产品编号:098683-01

一、本书定位

Java 是目前应用十分广泛的程序设计语言,它采用了面向对象程序设计技术,功能强大且简单易学,特别适用于 Internet 程序设计,已成为应用广泛的 JavaEE 应用开发的程序设计语言。JavaFX 是基于 Java 语言的下一代客户端平台和 GUI 框架,它提供了一个强大、流线化且灵活的框架,简化了现代的、视觉效果出色的 GUI 的创建。从 JavaFX 2.0 开始,JavaFX 开始完全用 Java 语言编写,并提供了一个 API。从 JDK 7 Update 4 开始,JavaFX 就已经与 Java 捆绑在一起了,并与 JDK 的版本号相一致。JavaFX 的提出是为了取代 Swing,但是现在仍然存在大量的 Swing 遗留代码,熟悉 Swing 编程的程序员也很多,所以 JavaFX 被定义为"未来的平台"。预计在未来的几年中,JavaFX 将会取代 Swing 并被应用到新的项目中,一些基于 Swing 的应用也会迁移到 JavaFX 平台。

NetBeans 是目前使用非常广泛、开源且免费的 Java 应用开发工具。作为 Oracle 公司官方认定的 Java 应用开发工具,NetBeans 的开发过程被认为最符合 Java 应用开发理念。

本书在编写上充分体现了简单易学的特点,步骤清晰,内容丰富,并配有大量插图,以帮助读者直观地理解基本内容,同时对内容的编排和示例的选择都做了严格控制,确保了一定的深度与广度。书中每个示例都配有执行结果插图,并对源代码进行了分析与讨论。本书采用 NetBeans IDE 作为 JavaFX 应用的开发与运行环境,该 IDE 可以从 Oracle 官网上免费下载和使用,实验环境的构建在单机与网络环境下都可以实现,具有软硬件环境投资少、经济实用、构建简单等特点。阅读本书的读者应该具有 Java 语言程序设计的基础,本书第 7 章涉及计算机图形学的相关概念与原理,读者可以参考选读。

二、本书特色

本书基于 JDK 9 与 NetBeans IDE 13 编写,除了介绍 JavaFX GUI 程序设

计的基本内容外,还精心选择并深入探讨了在 JavaFX 中具有代表性的 JavaFX Media、JavaFX 3D、JavaFX Web、JavaFX-Swing、JavaFX 图表、JavaFX 动画与视觉效果等应用开发技术。书中每章都有大量的从简单到综合的示例,同时对重点示例阐述了编程思想并归纳了必要的结论和概念。

  本书使用的计算机软件都可以通过 Internet 免费下载,即使读者的计算机没有与局域网或 Internet 相连接,也可以在一台独立的计算机上完成本书所有源代码的编译和运行。另外,本书的电子教案及源代码等配套资源均可在清华大学出版社网站上免费下载。

三、本书知识体系

  本书共 11 章,第 1 章介绍 JDK 9、NetBeans IDE 13 的下载、安装过程与基本结构,讲解基于 NetBeans 开发 Java 应用程序的基本原理与过程,并简要介绍 Oracle DB 11g XE 以及如何基于 NetBeans 连接与操作 Oracle DB 11g XE。第 2 章介绍 JavaFX 这个功能强大的新 GUI 框架,以及基于 NetBeans IDE 开发 JavaFX 应用程序的原理与方法。第 3 章介绍基于 NetBeans IDE 开发拥有图片与 TreeView 的 JavaFX 应用程序的方法。第 4 章介绍 JavaFX 的复选框 (CheckBox)、列表(ListView)和文本框(TextField)控件的用法。第 5 章介绍 JavaFX 菜单应用程序以及让 GUI 具有用户期望的外观的原理与方法。第 6 章通过一个实际的示例介绍 JavaFX Media 程序设计方面的知识。第 7 章介绍 JavaFX 的图形技术,包括 JavaFX 3D 图形入门、使用图像操作 API、使用 Canvas API。第 8 章介绍 JavaFX 嵌入式浏览器这个用户界面组件,其 API 提供 Web 查看器和浏览器的功能;介绍 JavaFX WebView 组件及其支持的 HTML5 功能;介绍如何将 WebView 组件添加到 JavaFX 应用程序的场景中,如何为当前文档运行特定的 JavaScript 命令,并将其加载到嵌入式浏览器中,如何从 JavaScript 调用 JavaFX 应用程序,如何使用 PopupFeatures 类为其设置其他 WebView 对象,并在单独窗口中打开文档,如何使用 WebHistory 类获取已访问页面的列表;讨论用于打印嵌入式浏览器的 HTML 内容的代码模式。第 9 章介绍基于 Swing 组件的 JavaFX 应用开发;探讨如何将 JavaFX 内容集成到 Swing 应用中,以及如何在 JavaFX 应用中使用 Swing 组件;通过若干综合示例介绍如何基于 Swing 组件进行 JavaFX 应用开发。第 10 章介绍 JavaFX 图表 (Chart)API 提供的方法,以及如何更改图表的外观、视觉和数据以使其成为一个易于扩展且灵活的 API。第 11 章介绍如何基于 JavaFX 开发具有变换、时间轴动画以及视觉效果的 JavaFX 应用,并基于示例介绍与它们相关的概念与实现原理。

本书由宋波编著,并负责书稿的修订、完善、统稿工作。本书从选题到立意,从酝酿到完稿,自始至终得到了学校、院系领导和同行教师,以及清华大学出版社相关老师的关心与指导,责任编辑为本书的出版工作付出了极大的辛苦与努力。本书也吸纳和借鉴了中外参考文献中的资料,在此一并致谢。

由于作者教学、科研任务繁重且水平有限,加之时间紧迫,对于书中存在的错误和不妥之处,诚挚地欢迎读者批评指正。

宋　波
2022 年 9 月

目录 CONTENTS

第1章 JavaFX 开发与运行环境 ………………………………………… 1
  1.1 JDK 的下载、安装与环境设置 ………………………………… 1
  1.2 NetBeans 的下载与安装 ………………………………………… 7
  1.3 NetBeans IDE 简介 ……………………………………………… 12
    1.3.1 NetBeans 菜单栏 ………………………………………… 13
    1.3.2 NetBeans 工具栏 ………………………………………… 14
    1.3.3 NetBeans 窗口 …………………………………………… 15
    1.3.4 代码编辑器 ……………………………………………… 21
  1.4 基于 IDE 开发 Java 应用 ……………………………………… 23
  1.5 Oracle DB XE 11g 简介 ………………………………………… 27
  1.6 Oracle DB XE 系统需求 ………………………………………… 27
  1.7 下载与安装 Oracle DB XE ……………………………………… 28
  1.8 Oracle XE DB 体系结构 ………………………………………… 31
    1.8.1 Oracle 实例 ……………………………………………… 31
    1.8.2 Oracle 数据库 …………………………………………… 33
  1.9 启动和停止 Oracle DB XE ……………………………………… 34
  1.10 连接 Oracle DB XE …………………………………………… 35
  1.11 Oracle Application Express …………………………………… 36
  1.12 基于 NetBeans 连接与操作 Oracle DB 11g XE ……………… 39
  1.13 小结 …………………………………………………………… 42

第2章 JavaFX GUI 编程概述 ……………………………………………… 43
  2.1 JavaFX 的基本概念 ……………………………………………… 43
  2.2 JavaFX 程序框架 ………………………………………………… 45

2.3　JavaFX 控件 Label ………………………………………………… 49

　2.4　JavaFX 控件 Button ………………………………………………… 51

　2.5　小结 …………………………………………………………………… 54

第 3 章　JavaFX 控件——Image、ImageView 与 TreeView ……………… 55

　3.1　Image 和 ImageView 控件 …………………………………………… 55

　3.2　TreeView 控件 ………………………………………………………… 57

　3.3　小结 …………………………………………………………………… 61

第 4 章　JavaFX 的其他控件 ………………………………………………… 62

　4.1　CheckBox ……………………………………………………………… 62

　4.2　ListView ……………………………………………………………… 65

　4.3　TextField ……………………………………………………………… 69

　4.4　小结 …………………………………………………………………… 71

第 5 章　JavaFX 菜单 ………………………………………………………… 72

　5.1　基础知识 ……………………………………………………………… 72

　5.2　MenuBar、Menu 和 MenuItem 概述 ………………………………… 73

　5.3　创建主菜单 …………………………………………………………… 75

　5.4　效果与变换 …………………………………………………………… 79

　5.5　小结 …………………………………………………………………… 84

第 6 章　JavaFX Media 应用开发 …………………………………………… 85

　6.1　JavaFX 支持的媒体编解码器 ………………………………………… 85

　6.2　HTTP 实时流媒体支持 ……………………………………………… 86

　6.3　创建 Media Player …………………………………………………… 86

　6.4　将媒体嵌入 Web Page ………………………………………………… 87

　6.5　创建 JavaFX 应用 …………………………………………………… 88

　6.6　控制媒体播放 ………………………………………………………… 90

　6.7　创建控件 ……………………………………………………………… 91

　6.8　添加逻辑功能代码 …………………………………………………… 94

　6.9　修改 EmbeddedMediaPlayer.java …………………………………… 98

　6.10　小结 ………………………………………………………………… 100

第 7 章　JavaFX 3D 应用开发 ………………………………… 101

- 7.1　Shape 3D ………………………………………………… 101
- 7.2　Camera 3D ……………………………………………… 103
- 7.3　SubScene ………………………………………………… 109
- 7.4　Light ……………………………………………………… 110
- 7.5　Material …………………………………………………… 112
- 7.6　Picking …………………………………………………… 113
- 7.7　构建 3D 示例应用程序 …………………………………… 115
- 7.8　Canvas …………………………………………………… 123
- 7.9　小结 ……………………………………………………… 133

第 8 章　JavaFX Web 应用开发 ……………………………… 134

- 8.1　JavaFX WebView 组件概述 ……………………………… 134
  - 8.1.1　WebEngine 类 ……………………………………… 135
  - 8.1.2　WebView 类 ………………………………………… 135
  - 8.1.3　PopupFeatures 类 ………………………………… 136
  - 8.1.4　其他特性 …………………………………………… 136
- 8.2　JavaFX 支持的 HTML5 功能 …………………………… 137
  - 8.2.1　Canvas 与 SVG …………………………………… 137
  - 8.2.2　媒体播放 …………………………………………… 137
  - 8.2.3　表单控制 …………………………………………… 138
- 8.3　历史记录维护 …………………………………………… 139
- 8.4　交互式元素标记 ………………………………………… 140
- 8.5　文档对象模型 …………………………………………… 141
- 8.6　Web Sockets …………………………………………… 141
- 8.7　Web Workers …………………………………………… 142
- 8.8　Web Font ………………………………………………… 142
- 8.9　将 WebView 组件添加到应用场景中 …………………… 143
- 8.10　创建工具栏 ……………………………………………… 144
- 8.11　调用 JavaScript 命令 …………………………………… 145
- 8.12　从 JavaScript 调用 JavaFX …………………………… 147
- 8.13　管理 Web 弹出窗口 ……………………………………… 148
- 8.14　获取访问页面列表 ……………………………………… 150

8.15 HTML 内容打印 ·············· 151
    8.15.1 使用打印 API ·············· 151
    8.15.2 添加上下文菜单以启用打印 ·············· 152
8.16 处理打印作业 ·············· 153
8.17 小结 ·············· 153

第 9 章 基于 Swing 组件的 JavaFX 应用开发 ·············· 155

9.1 JavaFX-Swing 的互操作性 ·············· 155
9.2 将 JavaFX 集成到 Swing 应用中 ·············· 156
    9.2.1 向 Swing 组件添加 JavaFX 内容 ·············· 156
    9.2.2 Swing-JavaFX 的互操作性与线程 ·············· 158
9.3 SimpleSwingBrowser 应用 ·············· 159
9.4 在 JavaFX 中实现一个 Swing 应用 ·············· 164
9.5 小结 ·············· 172

第 10 章 基于 JavaFX 的图表应用开发 ·············· 173

10.1 JavaFX 图表 API 的结构 ·············· 173
10.2 使用 JavaFX PieChart ·············· 174
10.3 使用 XYChart ·············· 182
10.4 改进示例的实现 ·············· 186
10.5 使用 LineChart ·············· 189
10.6 使用 BarChart ·············· 190
10.7 使用 StackedBarChart ·············· 192
10.8 使用 AreaChart ·············· 193
10.9 使用 StackedAreaChart ·············· 194
10.10 使用 BubbleChart ·············· 195
10.11 小结 ·············· 200

第 11 章 基于 JavaFX 开发动画与视觉效果 ·············· 201

11.1 在 JavaFX 中应用变换 ·············· 201
    11.1.1 变换概述 ·············· 201
    11.1.2 变换的类型与示例 ·············· 202
11.2 创建转换与时间轴动画 ·············· 206
    11.2.1 动画基础 ·············· 206

11.2.2　时间轴动画 …………………………………… 210
　　　11.2.3　树动画示例 …………………………………… 214
　11.3　创建视觉效果 …………………………………………… 226
　　　11.3.1　应用效果 ……………………………………… 227
　　　11.3.2　内部阴影效果 ………………………………… 233
　　　11.3.3　反射效果 ……………………………………… 234
　　　11.3.4　照明效果 ……………………………………… 235
　　　11.3.5　透视效果 ……………………………………… 237
　　　11.3.6　创建一个效应链 ……………………………… 238
　11.4　小结 ……………………………………………………… 241

附录1　图形教程的源代码 ……………………………………… 242

附录2　WebViewSample 应用的源代码文件 ………………… 243

附录3　示例源代码 ……………………………………………… 250

参考文献 …………………………………………………………… 258

Chapter 1
第1章 JavaFX 开发与运行环境

NetBeans 是目前使用广泛、开源且免费的 Java 应用开发工具。作为 Oracle 公司官方认定的 Java 应用开发工具，NetBeans 的开发过程被认为是最符合 Java 语言的开发理念。本章将介绍 JDK 9.0.4、NetBeans IDE 13.0 的下载、安装与环境设置；讲解如何基于 NetBeans IDE 开发 Java Application；简要介绍 Oracle DB 11g XE，以及如何基于 NetBeans IDE 连接与操作 Oracle DB 11g XE 等方面的知识。

1.1 JDK 的下载、安装与环境设置

在 Oracle 公司的网站 https://www.oracle.com/java/technologies/javase/javase9-archive-downloads.html 上可以免费下载 JDK 安装包，本书使用 JDK 9 版本（文件名为 jdk-9.0.4_windows-x64_bin.exe，64 位 OS）。在 Windows 10 OS 上安装 JDK 的具体操作步骤如下。

（1）关闭所有正在运行的程序，双击 Java SE 安装程序，进入安装向导界面，如图 1.1 所示。单击"下一步"按钮，进入更改文件夹界面，如图 1.2 所示。

（2）在图 1.2 所示的对话框中查找或创建将要安装的文件夹（本书为 E:\jdk-9），然后单击"确定"按钮，进入 JDK 9 安装进度界面，如图 1.3 所示。

（3）JDK 安装完成后，将会出现如图 1.4 所示的定制安装界面（安装 JRE 的界面）。单击"更改"按钮，将安装路径改为 E:\JRE-9（注意：要将 JDK 与 JRE 安装在两个不同的文件夹中），然后单击"下一步"按钮，出现如图 1.5 所示的安装界面。最后，安装完成后将出现如图 1.6 所示的界面。单击"关闭"按钮，即可完成 JDK 的安装工作。

JDK 9 成功安装之后，在指定的安装位置可以打开 JDK 9 的文件夹，如图 1.7 所示。

在 JDK 的安装文件夹下有 bin、include、lib 等子文件夹。下面是各个子文件夹的主要功能。

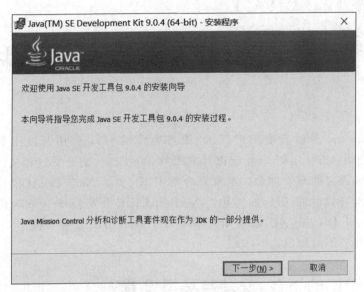

图 1.1　JDK 安装向导界面

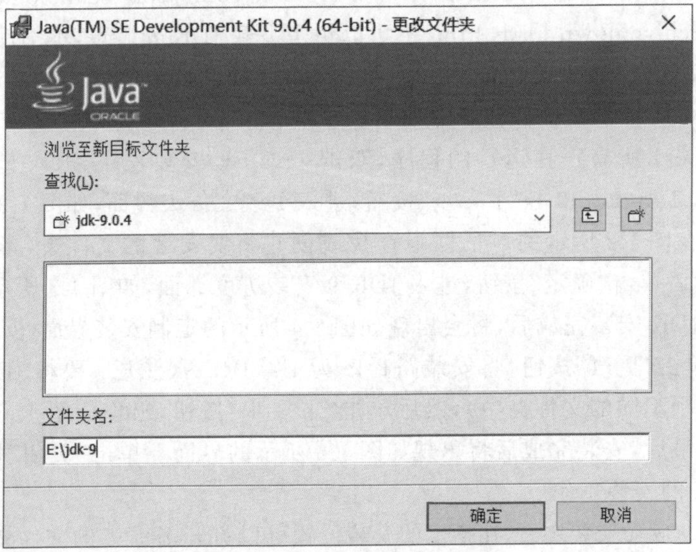

图 1.2　JDK 更改文件夹界面

图 1.3　JDK 安装进度界面

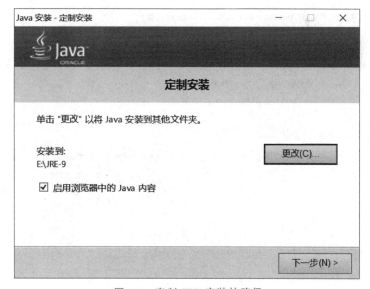

图 1.4　定制 JRE 安装的路径

图 1.5 JDK 与 JRE 安装界面

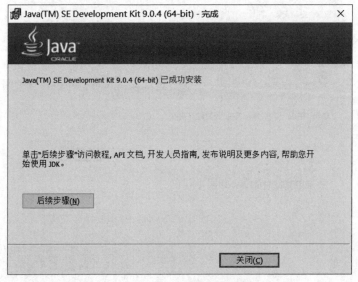

图 1.6 JDK 安装成功界面

第1章 JavaFX 开发与运行环境

图 1.7 JDK 9 目录结构及文件

- bin：用来存放开发 Java 程序所用的工具，例如编译指令 javac、执行指令 java 等。
- lib：用来存放开发工具包的类库文件。
- include：用来存放编译本地方法的 C++ 头文件。
- jre：安装在 E:\JRE-9 目录下，用来存放 Java 运行时的环境(JRE)。

> **注意**：如果要开发并运行 Java Application，则应当安装 JDK。安装 JDK 之后，也就包含了 JRE。如果只是运行 Java Application，则安装 JRE 就可以了。运行 Java Application 不仅需要 JVM，还需要类加载器、字节码检验器以及 Java 类库，而 JRE 恰好包含上述运行环境的支持。

编译和执行 Java Application 必须经过如下两个步骤：

第一步，将 Java 代码文件(扩展名为 java)编译成字节码文件(扩展名为 class)；

第二步，解释执行字节码文件。

实现以上两个步骤需要使用 javac 和 java 命令。通过以下操作步骤可以设置 Windows 10 OS 的环境变量并测试 JDK 的设置是否成功，才能正确地编译和执行 Java Application。

(1) 单击桌面上的"控制面板"图标，在弹出的对话框中单击"系统和安全"命令图标，显示"系统和安全"对话框。单击"系统"图标，在显示的对话框中单击"高级系统设置"项，在弹出的"系统属性"对话框中单击"环境变量"按钮，则将弹出"环境变量"对话框。单击"系统变量"选项组中的"新建"按钮，在弹出的"新建系统变量"对话框中输入变量名 Java_Home 和它的值 E:\jdk-9，单击"确定"按钮，如图 1.8 所示。

再新建一个 CLASSPATH 环境变量，其值为：E:\jdk-9\lib\dt.jar;E:\jdk-9\lib\tools.jar;E:\JavaExamples，如图 1.9 所示。

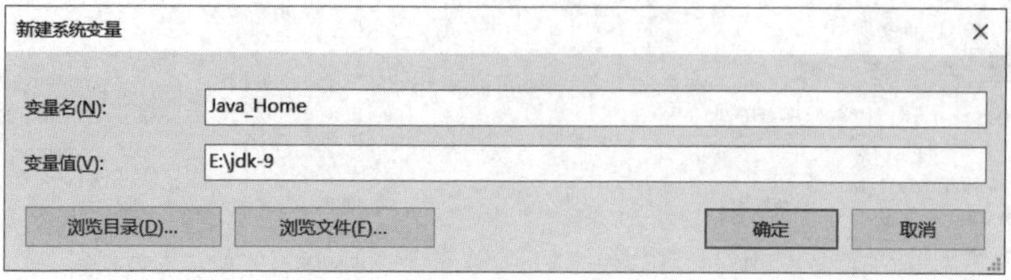

图 1.8 设置 Java_Home

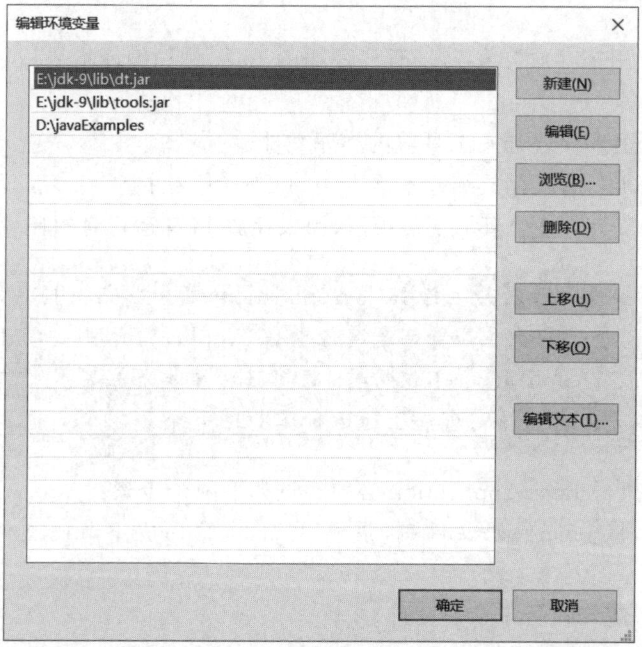

图 1.9 环境变量 CLASSPATH 的值

(2) 选择"系统变量"选项组列表框中的 PATH 变量，单击"编辑"按钮，在弹出的"编辑环境变量"对话框中为 PATH 变量添加 E:\jdk-9\bin 及 D:\JavaExamples，单击"确定"按钮，如图 1.10 所示。

通过上述操作的设置，Java 编译器命令 javac、Java 解释器命令 java 以及其他工具命令（例如 jar、appletviewer、javadoc 等）都将位于其安装路径下的 bin 目录中。

JDK 的安装和设置完成之后，就可以对 JDK 进行测试了。在命令行窗口中输入 java -version，按 Enter 键。如果系统显示输出如图 1.11 所示的 JDK 版本信息，则说明设置成功。

第 1 章　JavaFX 开发与运行环境

图 1.10　设置环境变量 PATH

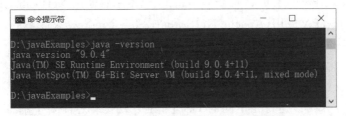

图 1.11　测试 Java 应用程序的编译及运行环境

1.2　NetBeans 的下载与安装

　　NetBeans 主要包括 IDE（集成开发环境）和 Platform（平台）两部分。其中，IDE 在平台基础上实现，并且平台本身也可以免费使用。NetBeans 可以运行在 Windows、Solaris 和 Mac 等 OS 上，可以开发标准的 Java Application、JavaFX 应用、Web 应用等。目前，NetBeans 的最新版本是 NetBeans 14.0。NetBeans 除了完全支持 Java SE、JavaEE、Java ME 和 JavaFX 以外，还新增了 JavaFX 编写器，能够以可视化的方式生成 JavaFX GUI 程序。其他的重要功能改进包括支持 PHP Zend 框架、Ruby on Rails 3.0，以及改进的 Java 编辑器、调试器和问题跟踪等。

- 代码编辑器：支持代码缩进、自动补全和高亮显示；可以自动分代码、自动匹配单词和括号、标注代码错误、显示和提示 JavaDoc；提供集成的代码重构、调试和 JUnit 单元测试。
- GUI 编辑器：在 IDE 中，可以通过拖曳设计基于组件的 GUI；IDE 内建有对本地化和国际化的支持，可以开发多种程序设计语言的应用程序。
- JavaEE 应用开发：支持 GlassFish、JBoss 以及 Tomcat 等 Web 服务器，支持 JavaEE 应用的开发。
- Web 应用开发：支持 Servlet/JSP、JSF、Struts、Ajax 和 JSTL 等 Web 技术的应用开发，提供编辑部署描述符的可视化编辑器以及调试 Web 应用的 HTTP 监视器，还支持可视化 JSF 开发。
- 协同开发：可以从官方网站免费下载 NetBeans Developer Collaboration，开发人员可以通过网络实时共享项目和文件。
- 支持可视化的手机程序开发，支持 Ruby 和 Rails 的开发，支持版本控制 CVS 和 Subversion。

1. 下载 NetBeans

目前，NetBeans IDE 的最新版本是 14.0，本书将使用 NetBeans IDE 13.0 开发与运行 JavaFX 应用（Apache-NetBeans-13-bin-windows-x64.exe）。可以从以下两个网站免费下载 Netbeans。

- http://www.oracle.com/technetwork/
- http://netbeans.org

NetBeans 可以运行在不同的 OS 上，在下载安装之前要了解 NetBeans 对系统的最低要求以及推荐配置。表 1.1 给出了 NetBeans 在 Windows OS 中的安装要求。

表 1.1 NetBeans 推荐系统配置

资源名称	最低要求	推荐配置
处理器	800MHz Intel Pentium 3 及以上	Intel Pentium 4 2.6GHz
内存	512MB	2GB
显示器	最小屏幕分辨率为 1024×768 像素	最小屏幕分辨率为 1024×768 像素
硬盘空间	750MB	1GB
Java SE	JDK 8 及以上版本	JDK 8 及以上版本

2. 安装 NetBeans

NetBeans 可以安装与运行在 Windows、Linux、Solaris 等 OS 上。本节以 Windows 10 OS 为目标平台,介绍 NetBeans 13.0 的安装方法和过程。安装之前需要安装 JDK 8 或以上版本。

(1) 双击安装文件 Apache-NetBeans-13-bin-windows-x64.exe,显示如图 1.12 所示的界面。

图 1.12　NetBeans 安装初始界面

(2) 单击 Customize 按钮,显示定制安装界面,如图 1.13 所示。

(3) 选择需要安装的功能,单击 OK 按钮,显示如图 1.14 所示的界面。

(4) 如图 1.14 所示,勾选 I accept the term in the license agreement 复选框。单击 Next 按钮,将打开如图 1.15 所示的对话框,该对话框用于设置安装路径。设置完成后,单击 Next 按钮,将打开如图 1.16 所示的界面。

(5) 单击 Install 按钮开始安装。安装完毕后,将显示如图 1.17 所示的安装成功界面。

单击 Finish 按钮完成 NetBeans IDE 的安装工作。安装程序将在 Windows 10 OS 的"开始"菜单中创建启动 IDE 的程序,并在桌面上创建用于启动 IDE 的图标。

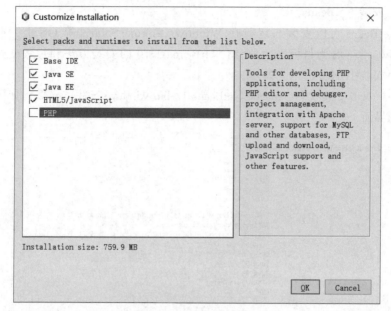

图 1.13　定制安装界面

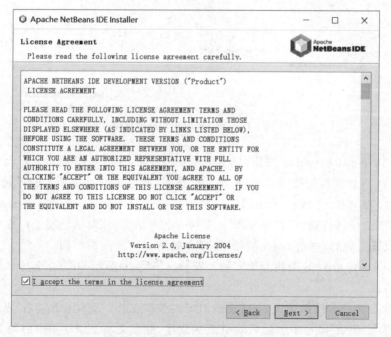

图 1.14　许可协议

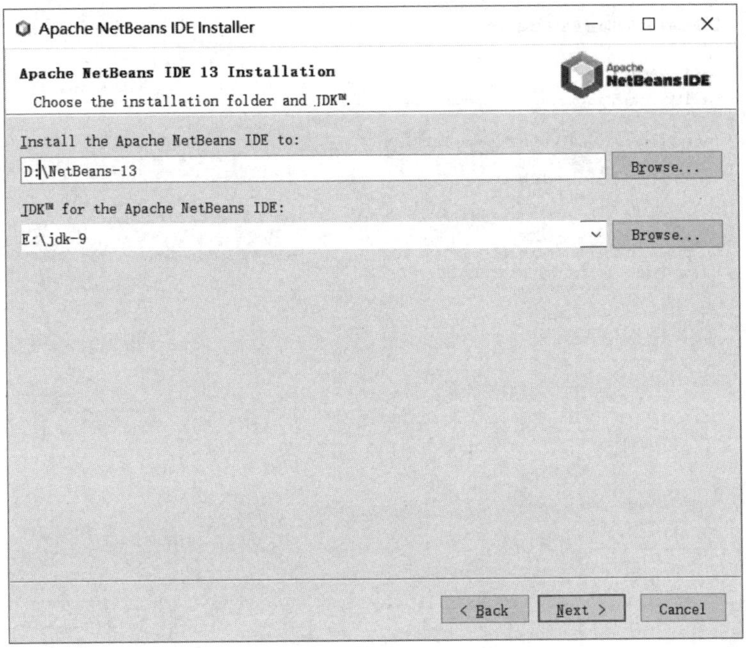

第 1 章　JavaFX 开发与运行环境　11

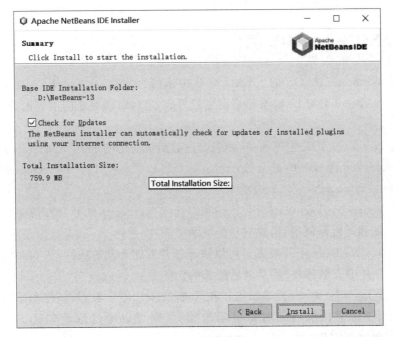

图 1.15　安装路径

图 1.16　安装摘要

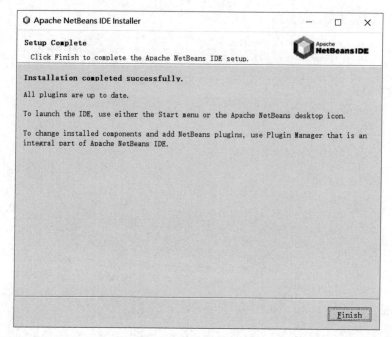

图 1.17 安装成功界面

1.3 NetBeans IDE 简介

在 Windows 10 OS 的"开始"菜单中选择"程序"→Apache NetBeans→Apache NetBeans IDE 13，将显示启动过程界面，之后启动完成的主界面如图 1.18 所示。

在 NetBeans IDE 主界面中，如果勾选起始页面中的 Show On Srartup 复选框，那么每次运行 IDE 时都会打开起始页。起始页包括 Learn & Discover、My NetBeans、What's New 3 个选项卡。

- Learn & Discover：开发人员可以访问 NetBeans 的开发和帮助文档，调试和运行示例项目，观看功能演示等。
- My NetBeans：开发人员可以快速打开近期开发的项目，从 NetBeans 更新中心安装插件、手动激活所需的功能等。
- What's New：开发人员可以在线浏览 NetBeans 教程、新闻和博客等。

如果不希望每次启动时都显示起始页，则可以通过取消勾选 Show On Srartup 复选框实现。

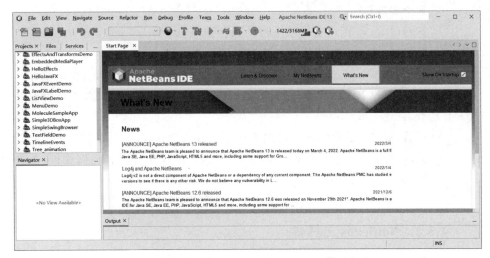

图 1.18　NetBeans 13.0 启动后的主界面

1.3.1　NetBeans 菜单栏

NetBeans 的菜单栏如图 1.19 所示,包括文件(File)、编辑(Edit)、视图(View)、导航(Navigate)、源(Source)、重构(Refactor)、运行(Run)、调试(Debug)、分析(Profile)、团队开发(Team)、工具(Tools)、窗口(Window)和帮助(Help)等子菜单。

图 1.19　NetBeans 12.0 主界面

- 文件(File):包括文件和项目的一些命令,例如新建、打开项目;打开、关闭文件;设置项目属性等。
- 编辑(Edit):包括复制、粘贴、剪切等各种简单的操作。
- 视图(View):包括各种视图的操作,可以控制工具栏中各个命令的显示/隐藏。
- 导航(Navigate):提供在编辑代码时进行跳转的各种功能,例如转至文件、上一个编辑位置、下一个书签等。
- 源(Source):提供对源代码的操作或控制,例如代码格式化、插入代码、修复代码、开启/关闭注释等。
- 重构(Refactor):提供对代码重新设定的功能,例如重命名、复制、移动、安全删除等。
- 运行(Run)：提供文件和项目的运行命令。
- 调试(Debug)：提供文件和项目的调试命令。
- 分析(Profile)：提供对内存使用情况或程序运行性能进行分析的命令。
- 团队开发(Team)：提供辅助团队开发的相关命令,例如团队开发服务器、创建生成作业等。
- 工具(Tools)：提供各种管理工具,例如库、服务器、组件面板等。
- 窗口(Window)：提供打开/关闭各种窗口的操作,例如项目、文件、服务器、导航、属性等。
- 帮助(Help)；提供有关 NetBeans 的帮助内容、联机文档等。

1.3.2 NetBeans 工具栏

NetBeans 工具栏如图 1.20 所示,它提供了诸如打开项目、复制和运行等一些常用的命令。把光标停留在某个按钮上将会显示该按钮的功能提示信息以及快捷键。

图 1.20 NetBeans 工具栏

开发人员可以通过以下两种方式对工具栏进行订制：
- 在"Toolbars(工具栏)"空白处右击,将弹出如图 1.21 所示的上下文菜单,选择执行 Customize 命令,可以在这里根据需要对工具条进行设置。

图 1.21 上下文菜单

- 打开菜单栏中的"视图"命令,选择"Toolbars(工具栏)"上下文菜单中的"Customize(定制)"选项,显示如图 1.22 所示的对话框。可以在该对话框中进行相关的订制操作。

图 1.22　订制工具栏

此外，开发人员可以通过选择工具栏上的 Performance 选项打开内存工具条，可以显示当前状态下的内存使用情况，如图 1.23 所示。

图 1.23　内存工具条

1.3.3　NetBeans 窗口

窗口是 NetBeans IDE 的重要组成部分，包括项目、文件、服务、属性、输出、导航等，每个窗口可用于实现不同的功能。

1. 项目窗口

项目窗口列出了当前打开的所有项目，是项目源的主入口。展开某个项目节点就会看到使用的项目内容的逻辑视图，如图 1.24 所示。项目是一个逻辑上的概念，容纳了一个应用程序的所有元素。一个项目可以包含一个文件，也可以包含多个文件。项目窗口可以包含一个项目，也可以包含多个项目。但是，在同一时刻只能有一个主项目。在项目窗口中可以进行项目的设置。项目窗口可以通过在菜单栏中选择"窗口"→"项目(J)"选项打开，或者通过快捷键 Ctrl+L 打开。一般地，一个项目可以包含以下逻辑内容。
- 源包：包括项目包含的源代码文件，双击某个源代码文件即可打开该文件并可在代码编辑器中进行编辑。
- 测试包：包含编写的单元测试代码。
- 库：包含该项目使用的库文件。
- 测试库：包含编写测试程序时使用的测试库。

右击项目窗口中的每个节点都会弹出相应的上下文快捷菜单，它包含所有主要的命令，如图 1.25 所示。

图 1.24 项目窗口

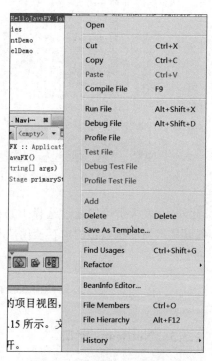

图 1.25 项目节点快捷菜单

2. 文件窗口

文件窗口显示基于目录的项目视图，包括项目窗口中未显示的文件和文件夹，以及支撑项目运行的配置文件，如图 1.26 所示。文件窗口可以通过菜单栏中的"窗口"→"文件（Files）"选项打开，或者通过快捷键 Ctrl+2 打开。

3. 服务窗口

服务窗口描述了 IDE 运行时资源的逻辑视图，包括数据库、Web 服务、服务器、团队开发服务器等,如图 1.27 所示。服务窗口可以通过"窗口"→"服务(Services)"选项打开,或者通过快捷键 Ctrl+5 打开。在服务窗口中,各节点的含义如下。

图 1.26　文件窗口

图 1.27　服务窗口

- 数据库(Databases):包括 Java DB 及其示例 sample、支持的数据库驱动程序,以及网络模式下的示例数据室库 sample。
- Web 服务(Web Services):用于管理所有相关的 Web 服务。
- 服务器(Servers):描述注册的所有服务器,包括 Apache Tomcat 和 Glass Fish Server。
- Maven 资源库(Maven Repositories):Apache Maven 是一种软件项目管理工具,提供了一个项目对象模型(POM)文件的概念以管理项目的构建,以及相关性和文档。
- 云(Cloud):云计算服务。
- Hudson 构建器(Hudson Builders):一个可扩展的持续集成引擎,用于持续、自动地构建/测试软件项目,以及监控一些定时执行的任务;在服务窗口中可以添加 Hudson 服务器。
- Docker:一个开源的应用容器引擎,基于 Go 语言并遵从 Apache 2.0 协议开源;Docker 可以让开发者打包它们的应用以及依赖包到一个轻量级、可移植的容器中,然后发布到任何流行的 Linux 机器上,也可以实现虚拟化。容器完全使用沙箱机制,相互之间不会有任何接口,更重要的是容器性能开销极低。Docker 从 17.03 版本之后分为 CE(Community Edition:社区版)和 EE(Enterprise Edition:企业版)。
- 任务资源库(Task Repositories):用于管理所有任务的资源库。

- Selenium 服务器(Selenium Server):Selenium 是一个用于 Web 应用测试的工具,Selenium 测试直接运行在浏览器中,就像真正的用户在操作一样;支持的浏览器包括 IE(7,8,9,10,11)、Mozilla Firefox、Safari、Google Chrome、Opera 等;Selenium 是一套完整的 Web 应用测试系统,包含测试的录制(selenium IDE)、编写及运行(Selenium Remote Control)和测试的并行处理(Selenium Grid);Selenium 的核心(Selenium Core)基于 JsUnit,完全由 JavaScript 编写,因此可以用于任何支持 JavaScript 的浏览器上。

4. 输出窗口

输出窗口用于显示来自于 IDE 的消息,消息种类包括调试程序、编译错误、输出语句、生成 Javadoc 文档等,如图 1.28 所示。输出窗口可以通过在菜单栏中选择"窗口"→"输出"选项打开,或者通过快捷键 Ctrl+4 打开。

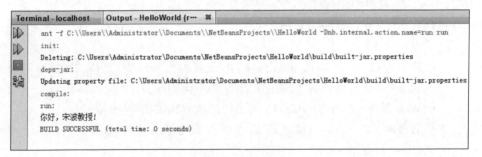

图 1.28 输出窗口

如果项目运行时需要输入信息,则输出窗口将显示一个新标签,并且光标将停留在标签处。此时,可以在窗口中输入信息,此信息与在命令行中输入的信息相同。

5. 导航窗口

导航窗口显示了当前选中文件包含的构造方法、成员方法、成员变量等信息,如图 1.29 所示。将光标停留在某成员的节点上,就可以显示 Javadoc 文档的内容。在导航窗口中,双击某成员节点可以在代码编辑器中直接定位该成员。在默认情形下,NetBeans IDE 的左下角显示导航窗口,可以通过在菜单栏中选择"窗口"→"导航"选项打开,或者通过快捷键 Ctrl+7 打开。

6. 组件面板窗口

组件面板管理器包含可以添加到 IDE 编译器中的各种组件。对于 Java 桌面应用程序,组件面板中的可用项包括容器、控件、窗口等,如图 1.30 所示。在该对话框中可以添加、删除组件面板窗口中的组件。

第 1 章　JavaFX 开发与运行环境　19

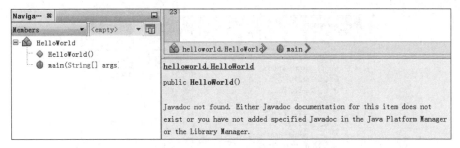

图 1.29　导航窗口

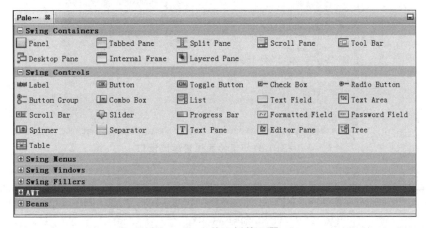

图 1.30　组件面板管理器

组件面板窗口可以通过在菜单栏中选择"窗口"→"组件面板"选项打开，或者通过快捷键 Ctrl+Shift+8 打开，如图 1.31 所示。

7. 属性窗口

属性窗口描述了项目包含的对象及对象元素具有的属性，开发人员可以在属性窗口中修改/查看这些属性。属性窗口显示了当前选定对象/组件的相关属性表单。图 1.32 左边为创建的 Java Appliction，右边描述了被选中组件的属性表单。

当单击图 1.32 中的 Find 按钮时，属性窗口则描述了该组件具有的属性、绑定表单、触发事件等。若要修改属性值，则可以单击属性值字段并直接输入新值，然后按 Enter 键即可。

如果属性值允许使用特定的值列表，则会出现下拉箭头，单击该箭头，然后选中值即可。如果属性编辑器适用于该属性，则会出现省略号(...)按钮，单击该按钮即可打开属性编辑器对属性值进行修改。

JavaFX 应用开发教程——基于 JDK 9 与 NetBeans 实现

图 1.31　组件面板管理器窗口

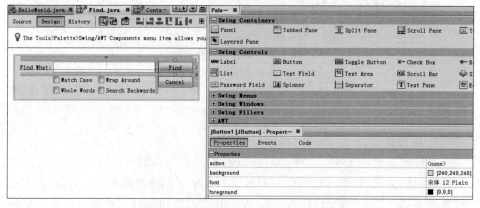

图 1.32　Java Application 与其组件属性

　　绑定表单描述了该组件与其他组件之间的关系，通过它可以修改绑定源及绑定表达式。事件表单列出了该选定控件支持的事件，通过触发相应的事件可以实现不同的功能，图 1.33 描述了 JButton 控件支持的 mouseClick（鼠标单击）事件。代码表单描述了被选定控件的相关代码，图 1.34 描述了 JButton 控件的代码。JButton1 在应用程序中的名称为 Find，该名称在程序中是唯一的，用来区分其他控件。

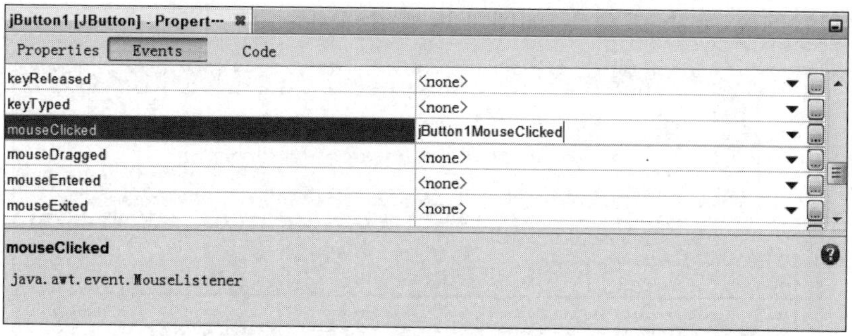

图 1.33 属性窗口事件表单

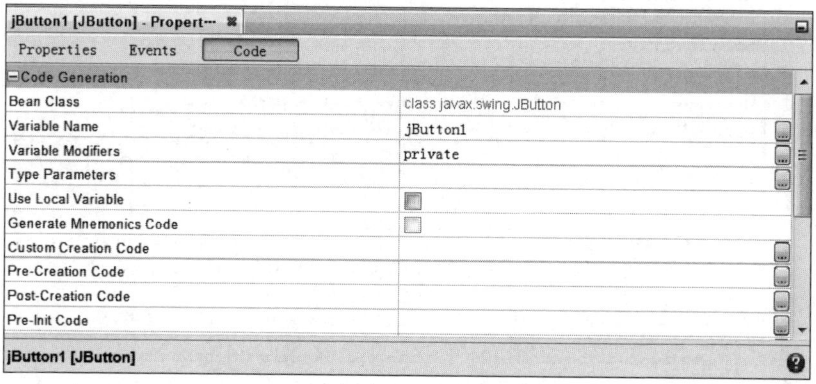

图 1.34 属性窗口代码表单

属性窗口可以通过在菜单栏中选择"窗口"→"属性"选项打开,或者通过快捷键 Ctrl+Shift+7 打开。

1.3.4 代码编辑器

代码编辑器提供了编写代码的场所,是 IDE 中使用最多的部分。代码编辑器提供了各种可以使代码编写更简单、更快捷的功能。

1. 代码模板

IDE 支持代码模板的功能。借助于代码模板,可以加快开发速度,积累开发经验,减少记忆与沟通成本。只要在源代码编辑器中输入代码模板的缩写,然后按 Tab 键或 Space 键,即可生成完整的代码片段。图 1.35 描述了定义的代码模板。

2. 快速编写代码

通过快速编写代码功能可以帮助用户快速查找并输入 Java 类名、表达式、方法名、组件名称、属性等。在输入字符之后,代码编辑器会自动显示提示菜单,列出包含的类、方法、变量等，如图 1.36 所示。

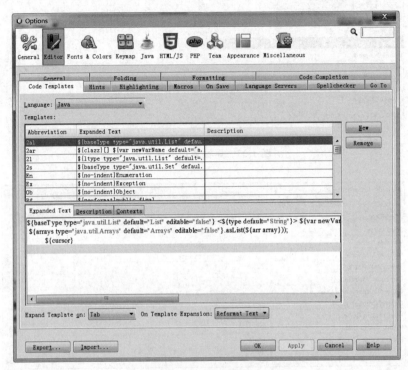

图 1.35 代码模板

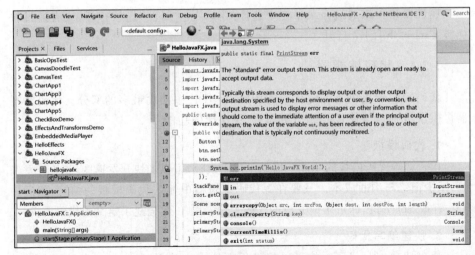

图 1.36 快速编写代码示意

1.4 基于 IDE 开发 Java 应用

用 NetBeans IDE 可以快速、便捷地开发 Java Application。在 IDE 中，所有的开发工作都基于"项目"完成。项目由一组源文件组成，即一个项目可以包含一个或一组源代码文件。此外，项目还包含用来生成、调试和运行这些源文件的配置文件。

用 IDE 生成的 Ant 脚本编译、调试和运行的项目称为标准项目。下面通过一个示例介绍创建 Java 标准项目的过程。该示例实现了一个银行账户类 basicAccount，可作为各种账户的基类。主类 bankAccount 用于输出账户的所有者信息和余额。假设 basicAccount 类具有下列成员。

- Owner：账户所有者。
- Balance：账户余额，一个只读的数值属性，该属性值取决于账户的存款额和取款额。
- Deposit：存款方法，该方法的参数为存款额，返回值为存款后账户的余额。
- Withdraw：取款方法，该方法的参数为取款额，返回值为取款后账户的余额。
- 构造方法：其参数为账户所有者的名称。

(1) 在 IDE 主菜单中选择"文件(File)"→"新建项目(New Project)"选项，打开"新建项目"对话框。在对话框的"Categeries(类别)"中选择 Java with Ant，在"Projects(项目)"中选择 Java Application 选项，如图 1.37 所示。

(2) 单击 Next 按钮，打开 New Java Application 对话框。在该对话框中，输入如图 1.38 所示的值，选择创建主类。单击 Finish 按钮，即可完成 Java 标准项目的创建，如图 1.39 所示。

此时，创建的标准项目包含主类 bankAccount，主类是一个项目的入口，并且一个 Java 标准项目只能有一个主类。图 1.40 所示为创建的项目的文件夹结构。IDE 将项目信息存储在项目文件夹和 nbproject 文件夹中，包括 Ant 生成的脚本、控制生成和运行的属性文件以及 XML 配置文件。源目录包含在项目文件夹中，名称为 src，test 目录用于保存项目的测试包。

主类用于输出账户所有者信息和余额，可以向 main( )方法中添加如下代码实现这个功能。

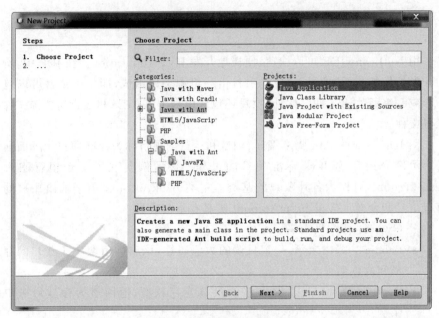

图 1.37 "新建项目"对话框

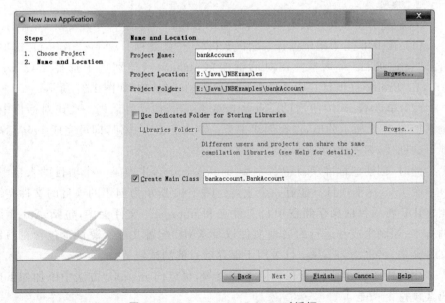

图 1.38 New Java Application 对话框

第 1 章　JavaFX 开发与运行环境　25

```
/*
 * To change this license header, choose License Headers in Project Properties.
 * To change this template file, choose Tools | Templates
 * and open the template in the editor.
 */
package bankaccount;
/**
 *
 * @author Administrator
 */
public class BankAccount {
    /**
     * @param args the command line arguments
     */
    public static void main(String[] args) {
        // TODO code application logic here
    }
}
```

图 1.39　创建的 Java 标准项目

图 1.40　项目目录结构

```
1.  package bankAccount;
2.  public class BankAccount {
3.      private String Owner="songbo";
4.      private double Balance=1000;
5.      public String getOwner() {
6.          return Owner;
7.      }
8.      public double getBalance() {
```

```
9.         return Balance;
10.    }
11.    public double Deposit(double amount) {
12.        Balance=Balance+amount;
13.        return Balance;
14.    }
15.    public double Withdraw(double amount) {
16.        Balance=Balance-amount;
17.        return Balance;
18.    }
19.    public static void main(String[ ] args) {
20.        BankAccount account=new BankAccount();
21.        System.out.println("账户所有者:"+account.getOwner());
22.      account.Deposit(100000.0);
23.      System.out.println("账户余额:"+account.getBalance());
24.    }
25. }
```

上述操作在创建 Java 标准项目的同时也创建了 Java 主类及 Java 包 bankAccount。如果没有勾选图 1.38 中的 Create Main Class 复选框，则需要另行创建 Java 包及 Java 主类。编译这个 Java 类并运行它，结果如下所示。

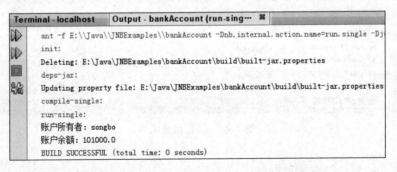

在 IDE 中，可以通过以下几种方式运行 Java 标准项目。

- 单击工具栏的"运行主项目"图标，该方法适用于运行主项目。若不是主项目，则可将其设置为主项目。
- 在项目窗口中选择要运行的项目并右击，选择"运行"选项即可，该方法适用于运行主项目和非主项目。
- 选择菜单项中的"运行"→"运行主项目"选项，该方法适用于运行主项目。

1.5 Oracle DB XE 11g 简介

Oracle DB XE 是 Oracle 公司于 2006 年 2 月推出的数据库新产品。Oracle DB XE 提供了针对 Windows OS 和 Linux OS 的产品版,开发人员可以借助已经得到证明、业界领先的强大基础架构开发和部署各种应用程序,然后在必要时可以升级到 Oracle DB 11g,而无须进行昂贵和复杂的移植。Oracle DB XE 可以安装在任意型号的计算机上,而计算机中 CPU 的数量是不受任何限制的。更为重要的是,这个世界上技术领先的数据库可以免费开发、部署和分发应用。另一方面,Oracle DB XE 在使用上也有一些限制,主要表现为:在主机上只能使用一个 CPU,最多存储 4GB 的用户数据,主机使用的最大内存为 1GB。

Oracle DB XE 是适用于以下人员的一款优秀的入门级数据库。

- 开发 Java、PHP、.Net、C/C++ 和开放式源代码应用的开发人员。
- 开发 SQL、PL/SQL 应用的开发人员。
- 需要用于培训和部署入门级数据库的 DBA。
- 希望获得可以免费分发的入门级数据库的独立软件供应者和硬件供应者。
- 在课程中需要免费数据库的教育机构和学生。

Oracle DB XE 包括以下 3 个产品。

- Oracle Database 11g Express Editor(Western European)——Oracle DB XE 英文版,该版本主要针对使用单字节拉丁语系的欧洲国家,安装文件名为 OracleXE.exe。
- Oracle Database 11g Express Editor(Universal)——Oracle DB XE 通用版,该版本主要针对使用双字节的国家,包括中国、日本、韩国等,安装文件名为 OracleXEUniv.exe。
- Oracle Database 11g Express Client——Oracle DB XE 客户端软件,适用于所有语言,安装文件名为 OracleXEClient.exe。

1.6 Oracle DB XE 系统需求

Oracle DB XE 可以运行在 Windows OS 和 Linux OS 上,本书以 Windows 10 Professional OS 为例介绍 Oracle DB XE 的内容。

Oracle DB XE 的软硬件系统的需求如下。

- System architecture——Intel(x86);

- Operating System——Windows XP Professional；Windows 7 Professional；Windows 10 Professional；
- Network——TCP/IP；
- Disk Space——1.2g；
- RAM——256MB minimum；
- Microsoft——MSI version 2.0 or later。

Oracle DB XE 可以使用浏览器作为控制台，从而实现各种数据库对象的访问和管理。Oracle DB XE 要求浏览器必须支持 JavaScript、HTML 4.0 和 CSS 1.0 标准。以下浏览器均可以作为 Oracle DB XE 的控制台。

- Microsoft Internet Explorer 6.0 or later；
- Netscape Communicator 7.2 or later；
- Mozilla 1.7 or later；
- Firefox 1.0 or later。

1.7 下载与安装 Oracle DB XE

Oracle 为 Oracle DB XE 创建了一个网站，免费注册会员并登录后就可以下载，网址为 https://www.oracle.com/database/technologies/xe-prior-release-downloads.html，文件名为 OracleXE 112_Win64.zip，如图 1.41 所示。

Oracle Database XE Prior Release Archive	
Oracle Database Express Edition (XE) Release 11.2.0.2.0 (11gR2).	
Name	Download
Oracle Database 11gR2 Express Edition for Windows x64	⬇ Download
Oracle Database 11gR2 Express Edition for Windows x32	⬇ Download
Oracle Database 11gR2 Express Edition for Linux x64	⬇ Download

图 1.41　Oracle DB XE 的下载

（1）在下载目录下，先将 OracleXE112_Win64.zip 解压缩，然后在解压缩目录下双击 setup.exe 文件，即可开始安装 Oracle DB XE，并显示安装界面，如图 1.42 所示。

（2）单击"下一步"按钮，则显示如图 1.43 所示的界面，勾选其下方的复选

第 1 章　JavaFX 开发与运行环境　29

图 1.42　Oracle DB XE 安装初始化

框(接受协议),然后单击"下一步"按钮,则显示确定安装目录界面(例如输入 E:\oraclexe),如图 1.44 所示。单击"下一步"按钮,则显示配置 SYS 和 SYSTEM 两个默认数据库账户的密码界面,如图 1.45 所示。输入口令,然后单击"下一步"按钮,则显示当前初始化配置信息界面,如图 1.46 所示。

图 1.43　接受协议界面

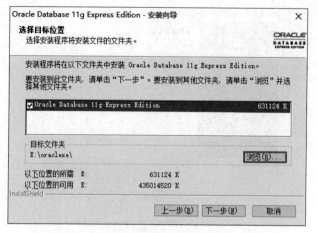

图 1.44 确定安装目录界面

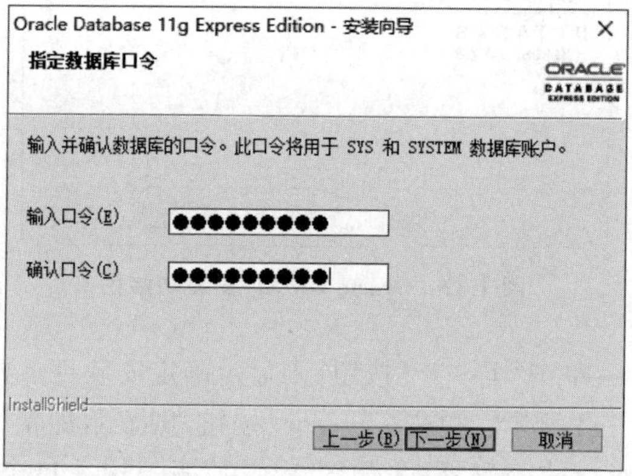

图 1.45 配置默认数据库账户密码

图 1.46 初始化配置信息界面

（3）单击"安装"按钮，则开始安装 Oracle DB XE，如图 1.47 所示。安装过程主要完成文件复制、启动服务和创建服务等工作。安装完成之后，将会显示完成界面。

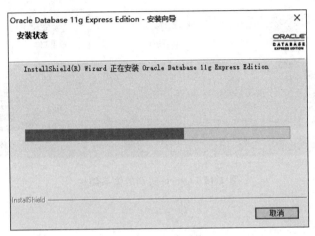

图 1.47　安装进程界面

1.8　Oracle XE DB 体系结构

Oracle XE DB 实际上指的是 Oracle 数据库管理系统，是一个管理数据库访问的计算机软件。Oracle DB XE 由 Oracle 数据库和 Oracle 实例两部分组成。

- Oracle 数据库——一个相关操作系统文件的集合，Oracle DB XE 用它存储和管理相关的信息。
- Oracle 实例——也称数据库服务或服务器，是一组 OS 进程和内存区域的集合，Oracle DB XE 用它管理数据库的访问。在启动一个与数据库文件关联的实例之前，用户还不能访问数据库。
- 一个 Oracle 实例只能访问一个 Oracle 数据库，而同一个 Oracle 数据库允许多个 Oracle 实例访问。

1.8.1　Oracle 实例

Oracle 实例由系统全局区（System Global Area，SGA）和程序全局区（Program Global Area，PGA）两部分组成，如图 1.48 所示。

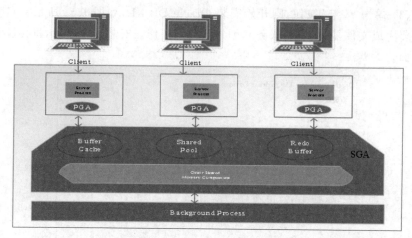

图 1.48　Oracle 实例的体系结构

1. 系统全局区

Oracle DB XE 中拥有以下进程。

- 用户进程——在客户机内存中运行的程序。例如,在客户机上运行的 SQL ＊Plus、企业管理器等都是用户进程。用户进程用于向服务器进程请求信息。
- 服务器进程——在服务器上运行的程序,接收用户进程发出的请求,并根据请求与数据库通信,完成与数据库的连接操作和 I/O 访问。
- 数据库后台支持进程——负责数据库的后台管理工作。

运行在客户机上的用户进程和运行在服务器上的服务器进程是同时进行的,OS 将为这些进程分配专用的内存区域以用于它们之间的通信,这个专用的内存区域称为 SGA(系统全局区)。

在 SGA 中,根据功能的不同划分为若干部分,比较重要的如下。

- Buffer Cache(高速缓冲区)——用于保存从数据文件中读取的数据区块副本,或用户已经处理过的数据。设立高速缓冲区的目的是减少访问数据时造成的磁盘读写操作,进而提高数据处理能力。所有的用户都可以共享高速缓冲区中的数据。
- Shared Pool(共享区)——当数据库接收到来自客户端的 SQL 语句后,系统将会解析 SQL 语句的语法是否正确。进行解析时,需要的系统信息以及解析后的结果都将保存在共享区。如果不同的用户执行相同的 SQL 语句,Oracle 实例则可以直接使用已经解析过的结果,这将大幅提高 SQL 语句的执行效率。

- 重置日志缓冲区（Redo Log Buffer）——用于记录数据库中所有数据修改的详细信息，这些信息的存储地点称为 Redo Entries，Oracle 实例将适时地将 Redo Entries 写入重置日志文件，以便数据库毁坏时可以进行必要的复原操作。

2. 程序全局区

Oracle 实例在运行时将创建服务进程以为用户的进程服务。对于每个来自客户端的请求，Oracle DB XE 都会创建一个服务进程以接收这个请求。PGA 是存储区中被单个用户进程使用的内存区域，是用户私有的，不能共享。PGA 主要用于处理 SQL 语句和控制用户登录等操作。

1.8.2 Oracle 数据库

Oracle 数据库是作为一个整体对待的数据集合，它由物理结构和逻辑结构两部分组成。物理结构是从数据库设计者的角度考察数据库的组成，而逻辑结构是从数据库使用者的角度考察数据库的组成。

1. 物理数据库结构

Oracle 数据库在物理上由数据文件（Data File）、重做日志文件（Redo Log Files）和控制文件 3 类系统文件组成。Oracle 数据库的这些文件为数据库信息提供了实际的物理存储。

- 数据文件——每个 Oracle 数据库都有一个以上的物理数据文件，它包括所有的数据库数据，像数据表和索引这样的具有逻辑数据库成分的数据，它们物理地存放在为数据库分配的数据文件中。
- 重做日志文件——每个 Oracle 数据库都拥有一个重做日志文件组。该组中含有 2 个以上的重做日志文件，这组重做日志文件称为数据库重做日志。重做日志由重做记录组成，每个记录都是一个描述数据库的单一基本更改的更改矢量组。重做日志的功能是记录对所有数据的修改。如果因为某种故障导致修改过的数据不能永久地写入数据文件，那么可以从重做日志获得相应的更改，使所做的工作不会丢失。
- 控制文件——每个 Oracle 数据库都拥有一个控制文件，它包含说明数据库物理结构的条目，例如数据库名、数据库的数据文件和重做日志文件的名称与位置、数据库的创建时间等。

2. 逻辑数据库结构

Oracle 数据库的逻辑成分包括表空间、模式对象、段、区、数据块，这些成分共同规定了数据库的物理表空间是如何利用的。
- 表空间——每个数据库至少有一个表空间,称为系统表空间。为了便于管理和提高运行效率,系统还自动创建了另外一些表空间。例如,用户表空间供一般用户使用,重做(UNDO)表空间供重做段使用。临时表空间供存放一些临时信息使用。一个表空间只能属于一个数据库。注意:一个表空间可以对应一个或多个数据文件,而一个数据文件只能属于一个表空间。
- 模式和模式对象——模式是数据库对象的集合,将模式中的数据库对象称为模式对象。模式对象是直接与数据库的数据有关的逻辑结构。例如,表、视图、序列、存储过程、同义词、索引等都是模式对象。一般地,一个模式对象对应于一个段,但在利用分区技术时也可以对应多个段。
- 数据块——Oracle 数据库中的数据是按照数据块存储的。数据块对应于磁盘上的物理数据库空间的一定数目的字节。在创建数据库时需要为数据库指定数据块的大小。数据库以数据块为单位使用和分配可用的数据库空间。
- 段——为某个逻辑结构分配的一组区。有以下一些不同类型的段。
    ① 数据段——每个表有一个数据段。所有表的数据都保存在其数据段的某个区中,对于分区表,每个分区有一个数据段。
    ② 索引段——每个索引有一个索引段,用于保存它的所有数据。对于分区索引,每个分区有一个索引段。
    ③ 回退段——管理员要为数据库创建一个或多个回退段,用于临时保存"撤销"信息,回退段中的信息用于生成一致性读取数据库的信息,在数据库恢复时,用于回退未提交的用户事务处理。
    ④ 临时段——在 SQL 语句需要临时工作区完成执行的工作时,将创建一个临时段。在该语句执行结束后,相应的临时段会返回系统,以备以后使用。

1.9 启动和停止 Oracle DB XE

Oracle DB XE 安装完毕后,它的数据库会自动启动。如果不进行设置,以后在启动 Windows OS 时,数据库也会自动启动。数据库启动后,将会占用系统的大量内存和 CPU 资源。如果不想让数据库自动启动,可以利用 Windows 的服务管理工具进行设置,设置方法如下:

- 选择"开始"→"设置"→"控制面板";
- 在"控制面板"中双击"管理工具",打开"管理工具";
- 在"管理工具"界面中双击"服务"图标;
- 在"服务"界面中找到 Oracle XE 实例并右击,则弹出一个上下文菜单,选择"属性"选项,将该服务实例的启动类型设置为"手动"。

Oracle DB XE 安装完毕后,会在"程序"菜单中生成启动、停止数据库等各种命令,如图 1.49 所示。如果将 OracleServiceXE 实例的启动类型设置为"手动",那么选择"启动数据库"命令则将启动数据库。图 1.50 所示为在命令行窗口显示的启动信息。

图 1.49　Oracle DB XE 命令菜单

图 1.50　命令行窗口的启动信息

1.10　连接 Oracle DB XE

Oracle DB XE 提供了一个命令行工具,用于实现与本地或远程数据库的连接。使用方法是选择"开始"→"程序"→Oracle Database 11g Express Editor→"运行 SQL 命令行",输入如图 1.51 所示的命令。

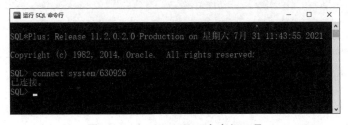

图 1.51　Oracle DB XE 命令行工具

1.11 Oracle Application Express

Oracle DB XE 使用浏览器作为控制台，利用这个可视化工具可以方便地实现 Oracle 数据库安全策略方面的管理。如图 1.52 所示，选择"入门"选项，就可以启动 Oracle DB XE 的控制台界面，如图 1.53 所示。

图 1.52　启动 Oracle DB XE 控制台

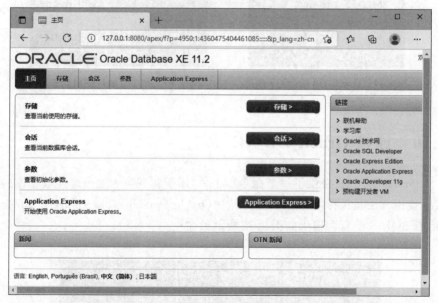

图 1.53　Oracle DB XE 控制台界面

单击 Application Express 按钮，则显示如图 1.54 所示的界面。输入用户名 SYS，密码 630926，单击 Login 按钮，则将显示如图 1.55 所示的界面，输入图中的用户名和密码（安装时设置的密码），单击 Login 按钮，则将进入如图 1.56 所示的 Application Express 窗口。

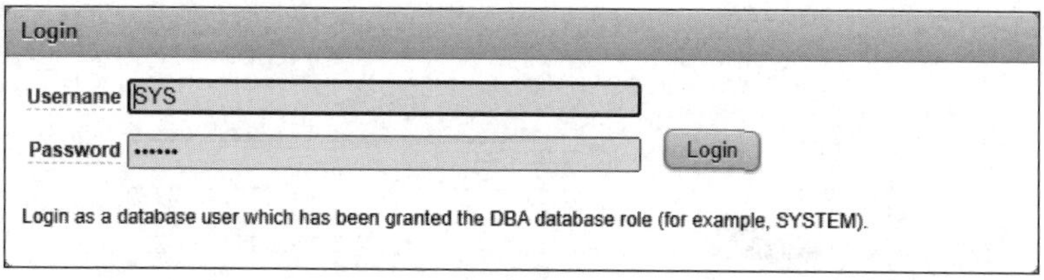

图 1.54 登录 Application Express

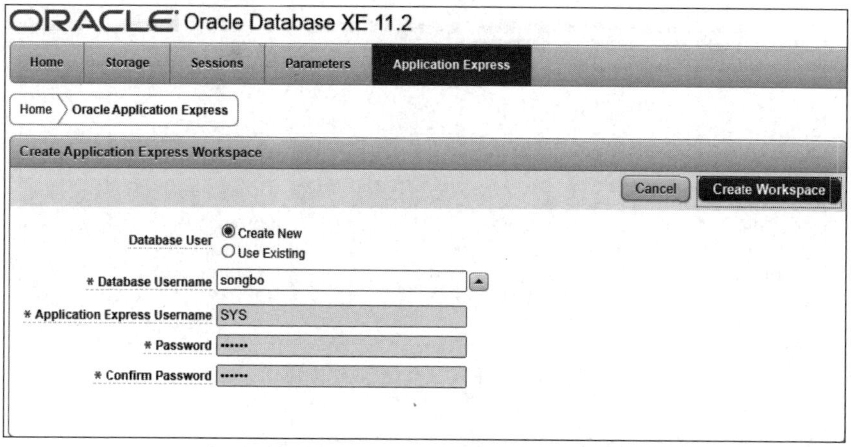

图 1.55 Application Express 窗口

输入图 1.54 所示的相应参数，单击 Create Workspace 按钮，则将显示如图 1.56 所示的界面。

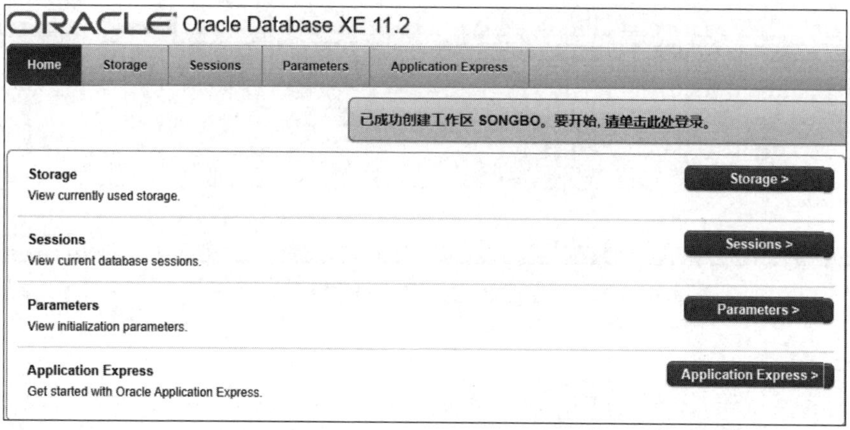

图 1.56 创建工作区 songbo

单击"请单击此处登录"链接,则将显示如图 1.57 所示的界面。

图 1.57　输入工作区和身份证明

单击"登录"按钮,则将显示如图 1.58 所示的界面。

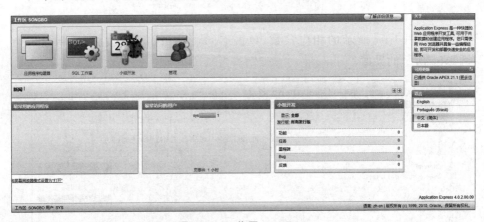

图 1.58　工作区 SONGBO

1.12 基于 NetBeans 连接与操作 Oracle DB 11g XE

（1）启动 NetBeans IDE，在菜单栏中选择 Services 选项，则将在浏览窗口出现 NetBeans 的服务窗口，如图 1.59 所示。其将显示 NetBeans 的各种服务选项，例如 JavaDB、Drivers、Server 等。

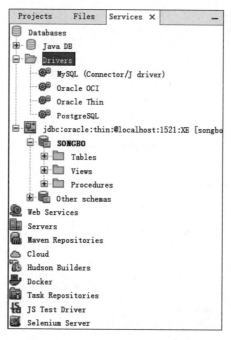

图 1.59　NetBeans IDE 的服务窗口

（2）从 Oracle 公司网站下载 ojdbc6.jar 驱动程序文件（本书的源代码压缩文件中也有提供），将其解压缩并保存在某一文件夹中。

（3）从 Windows 的"开始"菜单中启动 Oracle DB 11g XE 的数据库服务。

（4）展开 Drivers 节点，选择 Oracle Thin 节点并右击，选择自执行 Connect Using 选项，则将显示如图 1.60 所示的 New Connection Wizard 对话框。

（5）在图 1.61 所示的窗口中的 Driver 下拉列表中选择 Oracle Thin 节点。在 Driver File 区域单击 Add 按钮，选择 ojdbc6.jar 文件。

（6）单击 Next 按钮，则将进入如图 1.62 所示的窗口。

（7）单击 Test Connection 按钮，如果在对话框的底部显示 Connection Succeeded 的信息，则说明连接成功。

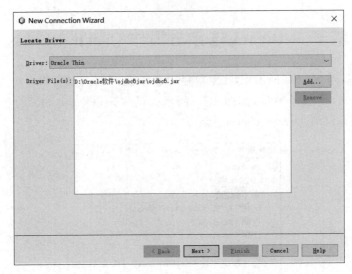

图 1.60　New Connection Wizard 对话框

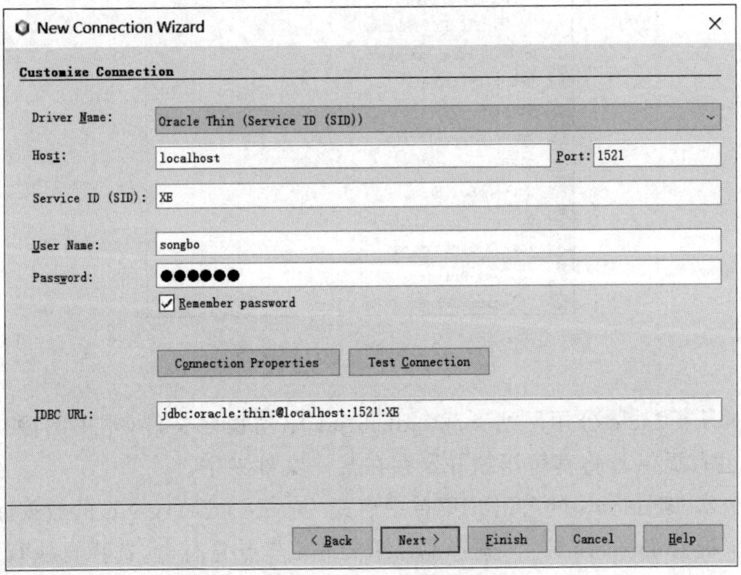

图 1.61　Customize Connection

　　(8) 单击 Next 按钮,则将显示如图 1.62 所示的界面。从 Select schema 下拉列表中选择 SONGBO 数据库模式。

　　在 NetBeans IDE 中设置连接 Oracle DB 11g XE 之后,就可以在 IDE 中进行各种数据库对象的操作了。如图 1.63 所示,展开创建的 jdbc:oracle:thin:

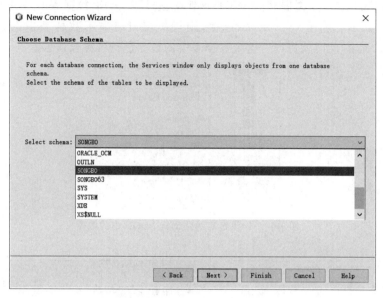

图 1.62　Choose Database Schema

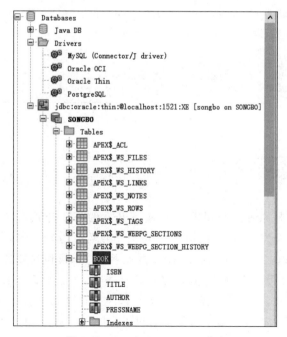

图 1.63　Oracle DB 11g XE 节点

@localhost…节点,再展开 SONGBO 节点,选择 BOOK 数据表节点并右击,如图 1.64 所示。选择 View Data 选项,则代码窗口将显示数据表中的数据,如图 1.65 所示,此时就可以对这个数据表中的数据进行各种操作了。

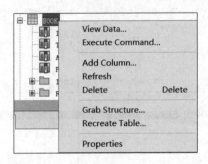

图 1.64　显示选定的数据表 BOOK 中的数据

#	ISBN	TITLE	AUTHOR	PRESSNAME
1	1-302-0101-X/TP	Web应用高级教程	GregBarish	清华大学出版社
2	3-4032-0306-X/TP	Java应用开发教程	宋波	电子工业出版社
3	5-503-0506-X/XL	Java应用设计	宋波	人民邮电出版社
4	4-402-5011-T/XT	Oracle 9i数据库高级管理	飞思科技	电子工业出版社

图 1.65　数据表 BOOK 中的数据

1.13　小结

　　NetBeans 是原 Sun 公司用 Java 语言开发的一个开源工具,是 Sun 公司官方认定的 Java 开发工具(现已被 Oracle 公司收购)。本章介绍了 NetBeans IDE 的一些基础知识,包括 NetBeans 13.0 的新特性、下载和安装方法、IDE 各个主要部分的简介、用 IDE 开发 Java Application 的方法、基于 IDE 连接与操作 Oracle DB 11g XE 等。NetBeans 是功能非常强大的 Java 开发工具。本章只进行概括性的介绍,想进一步学习的读者可以参阅 NetBeans 的帮助文档。

第 2 章 JavaFX GUI 编程概述

Java 语言最初的 GUI(Graphics User Interface)框架是 AWT，后来又开发了 Swing——提供了创建 GUI 的一种更可行的方法。为了更好地满足现代 GUI 的需求以及 GUI 设计的改进，Java 语言提出了下一代 GUI 框架——JavaFX。本章将简要介绍这个功能强大的新框架，以及基于 NetBeans IDE 开发 JavaFX 应用程序的原理与方法。

2.1 JavaFX 的基本概念

与所有成功的计算机编程语言一样，Java 语言也在不断地演化和改进。演化过程中重要的表现之一就是 GUI 框架。Java Swing 提供了创建 GUI 的一种良好的方法，并取得了巨大的成功，它一直是 Java 语言中主要采用的 GUI 框架。如今，消费类应用，特别是移动应用变得越来越重要，而这类应用要求 GUI 具有令人振奋的视觉效果。为了更好地处理这类 GUI，促使了 JavaFX 的问世。

JavaFX 是 Java 语言的下一代客户端平台和 GUI 框架。JavaFX 提供了一个强大、流线化且灵活的框架，简化了现代的、视觉效果出色的 GUI 的创建。JavaFX 的诞生分为两个阶段。最初的 JavaFX 基于一种称为 JavaFX Script 的脚本语言。但是，JavaFX Script 已经被弃用。从 JavaFX 2.0 开始，JavaFX 开始完全用 Java 语言编写，并提供了一个 API。从 JDK 7 Update 4 开始，JavaFX 就已经与 Java 捆绑在一起，并与 JDK 的版本号一致。JavaFX 的提出是为了取代 Swing。但是现在仍然存在大量的 Swing 遗留代码，并且熟悉 Swing 编程的程序员有很多，所以 JavaFX 被定义为未来的平台。预计在未来的几年，JavaFX 将会取代 Swing 应用到新的项目中，一些基于 Swing 的应用也会迁移到 JavaFX 平台上。

1. JavaFX 包

JavaFX 组件是轻量级的，并以一种易于管理、直接的方式处理事件。

JavaFX 的元素包含在以 javafx 为前缀的包中。从 JDK 9 开始,JavaFX 包都组织到了模块中,例如 javafxbase、javafx.graphics 和 javafx.controls 等。

2. stage 与 scene

JavaFX 使用的核心比喻是舞台(Stage)。正如现实中的舞台表演,舞台是有场景(Scene)的。也就是说,舞台定义了一个空间,场景定义了在空间中发生了什么。用专业术语讲,舞台是场景的容器,场景是其组成元素的容器。因此,所有 JavaFX 程序都有至少一个舞台和一个场景。这些元素封装在 Stage 和 Scene 这两个类中。Stage 是一个顶级容器,所有的 JavaFX 程序都能够自动地访问一个 Stage,称为主舞台(Primary Stage)。当 JavaFX 程序启动时,JRE 将会提供主舞台。尽管也可以创建其他舞台,但是对许多程序而言,主舞台是唯一需要的舞台。简而言之,Scene 是组成场景中的元素的容器,这些元素包括控件(例如按钮、标签和复选框)、文本和图形。为了创建场景,需要把这些元素添加到 Scene 容器的实例对象中。

3. 节点与场景图

场景中的单独元素叫作节点(Node)。例如,命令按钮就是一个节点。节点可以由一组节点组成。节点也可以有子节点。具有子节点的节点叫作父节点(Parent Node)或分支节点(Branch Node)。没有子节点的节点叫作终端节点或叶子(Leave)。场景中所有节点的集合创建出了场景图(Scene Graph),场景图构成了树(Tree)。场景图中有一种称为根节点(Root Node)的顶级结点,是场景图中唯一没有父节点的节点。也就是说,除了父节点以外,其他所有的节点都有父节点,而且所有节点都或直接或间接地派生自根节点。所有节点的基类都是 Node。有一些类直接或间接地派生自 Node 类,包括 Parent、Group、Region 和 Control 等。

4. 布局

JavaFX 提供的布局窗格用于管理在场景中放置元素的过程。例如,FlowPane 类提供了流式布局,GirdPane 类提供支持基于网格的行/列布局。布局窗格类包含在 javafx.scene.layout 包中。

5. Application 类和生命周期方法

JavaFX 程序必须是 javafx.application 包中的 Application 类的子类。因此,用户应用程序类将扩展 Application。Application 类定义了 3 个可被重写的生命周期方法,分别是 init( )、start( ) 和 stop( ) 方法。

- voidinit( )——程序开始执行时将调用该方法,用于执行各种初始化工作。但是它不能用于创建舞台或构建场景。如果不需要进行初始化,则不需要重写这个方法,因为系统会默认提供一个空版本的 init( )方法。
- abstract void start(StageprimaryStare)——该方法在 init( )方法之后被调用,是程序开始执行的地方,可以用来构造和设置场景。它接受一个 Stage 对象的引用作为参数。这是由运行时系统提供的舞台(主舞台),它是抽象方法,程序必须重写这个方法。
- void stop( )——程序终止时将调用该方法。该方法可以执行清理和关闭工作。如果不需要执行这些操作,可以使用默认的空版本。
- public static void launch(String ... args)——为了启动一个独立的 JavaFX 程序,必须调用该方法。其中,args 是一个指定了命令行实参的字符串列表,可以为空。调用 launch( )方法将开始构造程序,之后调用 init( )和 start( )方法。直到程序终止,该方法才会返回。

2.2 JavaFX 程序框架

所有的 JavaFX 程序都具有相同的基本框架。下面的示例演示了如何启动程序以及生命周期方法何时被调用。

启动 NetBeans IDE 13.0,选择 File→New Project 选项,则将显示如图 2.1 所示的对话框。在 Categories 区域中选择 Java with Ant→JavaFX 选项,在

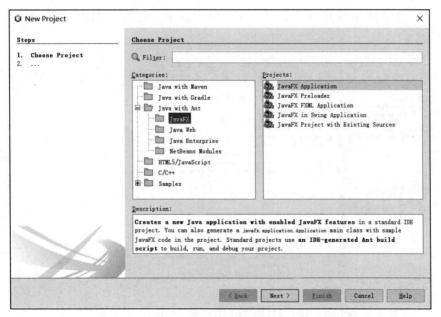

图 2.1 New Project 对话框

projects 区域选择 JavaFX Application 选项，然后单击 Next 按钮，则显示如图 2.2 所示的对话框。该对话框主要定义 JavaFX 应用的名称与保存路径。

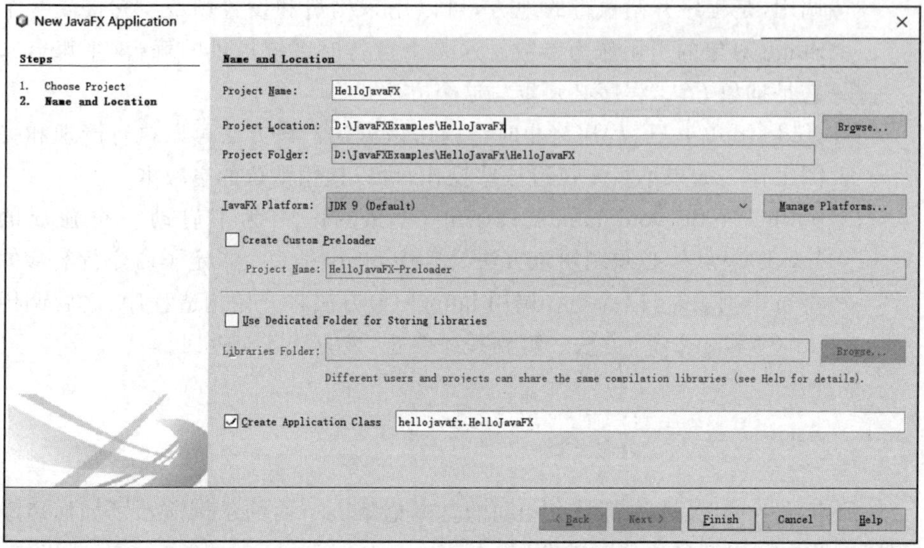

图 2.2 JavaFX Application 的 Name and Location 定义对话框

单击 Finish 按钮，完成 JavaFX 应用的向导定义，同时将在代码窗口、浏览器窗口以及浏览器结构窗口生成对应的 JavaFX 应用的相关信息，如图 2.3 所示。

图 2.3 NetBeans IDE 生成的 JavaFX 应用的代码、工程、代码结构等信息

在代码窗口中,将 IDE 生成的源代码修改成例 2.1 所示的代码,选择 Run→Run Project(HelloJavaFX)选项执行这个程序,执行结果如图 2.4 所示。

图 2.4　HelloJavaFX 程序的执行结果

【例 2.1】　HelloJavaFX.java 实例程序。

```
1.  package hellojavafx;
2.  import javafx.application.Application;
3.  import javafx.event.ActionEvent;
4.  import javafx.event.EventHandler;
5.  import javafx.scene.Scene;
6.  import javafx.scene.control.Button;
7.  import javafx.scene.layout.StackPane;
8.  import javafx.stage.Stage;
9.  /**
10.  * @author SongBo
11.  */
12. public class HelloJavaFX extends Application {
13.     @Override
14.     public void start(Stage primaryStage) {
15.         Button btn = new Button();
16.         btn.setText("Say 'Hello JavaFX World!'");
17.         btn.setOnAction(new EventHandler<ActionEvent>() {
18.             @Override
19.             public void handle(ActionEvent event) {
20.                 System.out.println("Hello JavaFX World!");
21.             }
22.         });
```

```
23.        StackPane root = new StackPane();              //创建根节点
24.        root.getChildren().add(btn);
25.        Scene scene = new Scene(root, 300, 250);        //创建场景
26.        primaryStage.setTitle("Hello JavaFX World!");
27.        primaryStage.setScene(scene);                   //设置舞台场景
28.        primaryStage.show();                            //显示场景
29.    }
30.    /**
31.     * @param args the command line arguments
32.     */
33.    public static void main(String[] args) {
34.        launch(args);
35.    }
36. }
```

【执行结果】

单击 Say 'Hello JavaFX World!'按钮,则将在 IDE 的程序输出窗口显示执行结果信息:

```
Hello JavaFX World!
```

【分析讨论】

- 一般地,在 JavaFX 程序中,生命周期方法不会像 System.out 那样输出任何信息,这里只是为了演示。
- 只有 JavaFX 程序必须执行特殊的启动和关闭操作时,才需要重写 init( )和 stop( )方法。否则,可以使用 Application 类为这些方法提供的默认实现。
- 该程序导入了 7 个包。其中,比较重要的有 4 个包:javafx.application 包——包含 Application 类;javafx.scene 包——包含 Scene 类;javafx.stage 包——包含 Stage 类;javafxsence.layout 包——提供 StackPane 布局窗格。
- 第 12 句创建了程序类 HelloJavaFX,它扩展了 Application 类。所有的 JavaFX 程序都必须派生自 Application 类。
- 程序启动后,JavaFX 运行时系统首先调用 init( )方法。当然,如果不需要初始化,就不必重写 init( )方法。注意:init( )方法不能用于创建 GUI 的舞台和场景部分,它们由 start( )方法构造和显示。
- 当 init( )方法完成后,star( )方法开始执行。在这里创建最初的场景,

并将其设置给主舞台。
- ◆ 首先,start( )方法有一个 Stage 类型的形式参数。调用 start( )方法时,这个形式参数将接受对程序主舞台的引用,程序的场景将被设置给这个舞台。
- 第 16 句创建了一个按钮对象 btn,并设置了这个按钮的标题。
- 第 23 句为场景创建了根节点。根节点是场景图中唯一没有父节点的节点。
- 第 24 句把按钮添加到了根节点中;第 25 句用根节点构造了一个场景(Scene)实例对象,并指定了宽度和高度;第 26 句设置了场景的标题;第 27 句把场景设置为了舞台的场景;第 28 句把场景在舞台上显示了出来。也就是说,show( )方法显示了舞台创建的按钮和场景。
- 当关闭程序时,JavaFX 将会调用 stop( )方法将窗口从屏幕上移除。另外,如果程序不需要处理任何关闭动作,就不必重写 stop( )方法,因为它有系统的默认实现。

2.3　JavaFX 控件 Label

JavaFX 提供了一组丰富的控件,标签(Label)是最简单的控件。标签是 JavaFX 的 Label 类的实例对象,包含在 javafx.secne.control 包中,继承了 Labeled 和 Control 等几个类。Labeled 类定义了带标签元素(包含文本的元素)共有的一些特性,Control 类定义了与所有控件相关的特性。Label 的构造方法如下。

Label(String str)——str 是要显示的字符串。

- 创建标签后,必须把它添加到场景的内容中,即把该控件添加到场景图中。

  要对场景图的根节点调用 getChildren( )方法,该方法返回一个 ObservableList<Node>形式的子节点列表。ObservableList 在 javafx.collections 包中定义,并继承了 java.util.List,它是 Java 语言的 Collections Framework 的一部分。只需要对 getChildren( )方法返回的子节点列表调用 add( )方法,并传入添加节点(标签)的引用即可。

【例 2.2】创建一个显示标签的 JavaFX 程序。

```
1.  package javafxlabeldemo;
2.  import javafx.application.*;
```

```
3.   import javafx.scene.*;
4.   import javafx.stage.*;
5.   import javafx.scene.layout.*;
6.   import javafx.scene.control.*;
7.   public class JavaFXLabelDemo extends Application {
8.     public static void main(String[] args) {
9.       //通过调用 launch()方法启动程序
10.      launch(args);
11.    }
12.    //重写 start()方法
13.    public void start(Stage myStage) {
14.      //设置 stage 的标题
15.      myStage.setTitle("Use a JavaFX label.");
16.      //使用 FlowPane 作为根节点的布局
17.      FlowPane rootNode = new FlowPane();
18.      //创建场景
19.      Scene myScene = new Scene(rootNode, 300, 200);
20.      //设置舞台的场景
21.      myStage.setScene(myScene);
22.      //创建标签
23.      Label myLabel = new Label("JavaFX is a powerful GUI");
24.      //把标签添加到根节点(舞台)
25.      rootNode.getChildren().add(myLabel);
26.      //在舞台上显示这个场景
27.      myStage.show();
28.    }
29. }
```

【执行结果】

执行结果如图 2.5 所示。

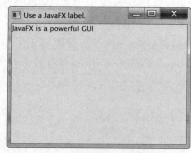

图 2.5　程序执行结果

【分析讨论】
- 需要注意的是第 25 句,其功能是把标签添加到 rootNode 的子节点列表中。
- ObservableList 类也提供了 addAll( )方法,它可以在一次调用中把两个或多个子节点添加到场景中。
- 如果要从场景中删除节点,则可以使用 remove( )方法。例如:

```
rootNode.getChildren( ).remove(myLabel);
```

上述代码从场景中删除了 myLabel。

2.4 JavaFX 控件 Button

在 JavaFX 程序中,事件处理十分重要,因为多数 GUI 控件都会产生事件,然后由程序处理这些事件。按钮事件是常用的处理事件的类型之一。本节将介绍按钮及其事件处理。

1. 事件处理基础

JavaFX 事件的基类是 javafx.event 包中的 Event 类。Event 类继承了 java.util.EventObject,JavaFX 事件与其他 Java 事件共享相同的基本功能。Event 类有几个子类,这里使用的是 ActionEvent 类,它用来处理按钮生成的动作事件。JavaFX 为事件处理使用了委托事件模型方法。

- 为处理事件,首先必须注册一个处理程序,作为该事件的监听器。
- 事件发生时将调用监听器。监听器必须响应该事件,然后返回。
- 事件是通过实现 EventHandler 接口处理的,该接口包含在 javafx.event 包中,是一个泛型接口。其语法形式为 Interface EventHandler＜T extends Event＞,T 指定了处理程序将要处理的事件类型。该接口定义了 handler( )方法,接受事件对象作为形式参数,如下所示。

```
void handler(T eventObj)
```

eventObj 是产生的事件。事件处理程序通过匿名内部类或 lambda 表达式实现,也可以通过使用独立的类实现(例如,事件处理程序需要处理来自多个事件源的事件)。

2. 按钮控件

在 JavaFX 中,命令按钮由 javafx.scene.control 包中的 Button 类提供。

Button 类继承了很多基类——ButtonBase、Labeled、Regin、Control、Parent 和 Node 等。按钮可以包含文件、图形或两者兼有。本节使用的按钮 Button 的构造方法如下：

```
Button(String str)
```

- 其中，str 是按钮中显示的信息。单击按钮时将产生 ActionEvent 事件。ActionEvent 包含在 javafx.event 包中。可以通过调用 getOnAction( )方法为该事件注册监听器。该方法的语法形式如下：

```
Final void setOnAction(EventHandler<ActionEvent> handler)
```

Handler 是事件处理程序。事件处理程序通过匿名内部类或 lambda 表达式实现。

setOnAction( )方法用于设置属性 onAction，该属性存储了对处理程序的引用。

- 事件处理程序将响应事件，然后返回。如果处理程序花费了太多时间，则将降低程序速度。对于耗时的操作，提倡使用独立的执行线程。

【例 2.3】 下面的程序使用了两个按钮和一个标签。这两个按钮的名称为 Up 和 Down。每次单击一个按钮时，标签就显示被按下的按钮是哪一个。

```
1.  package javafxeventdemo;
2.  import javafx.application.*;
3.  import javafx.scene.*;
4.  import javafx.stage.*;
5.  import javafx.scene.layout.*;
6.  import javafx.scene.control.*;
7.  import javafx.event.*;
8.  import javafx.geometry.*;
9.  public class JavaFXEventDemo extends Application {
10.     Label response;
11.     public static void main(String[ ] args) {
12.         //通过调用 launch()方法启动 JavaFX 应用
13.         launch(args);
14.     }
15.     //重写 start()方法
16.     public void start(Stage myStage) {
17.         //设置舞台标题
```

```
18.         myStage.setTitle("Use Platform.exit().");
19.         //使用 FlowPane 布局
20.         FlowPane rootNode = new FlowPane(10, 10);
21.         //Center the controls in the scene.
22.         rootNode.setAlignment(Pos.CENTER);
23.         //创建场景
24.         Scene myScene = new Scene(rootNode, 300, 100);
25.         //在舞台中设置场景
26.         myStage.setScene(myScene);
27.         //创建标签
28.         response = new Label("Push a Button");
29.         //创建按钮
30.         Button btnRun = new Button("Run");
31.         Button btnExit = new Button("Exit");
32.         //处理按钮 Run 事件
33.          btnRun.setOnAction((ae) -> response.setText("You pressed Run."));
34.         //处理按钮 Exit 事件
35.         btnExit.setOnAction((ae) -> Platform.exit());
36.         //在场景中添加标签和按钮
37.             rootNode.getChildren().addAll(btnRun, btnExit, response);
38.         //显示舞台的场景
39.         myStage.show();
40.     }
41. }
```

【执行结果】

程序执行结果如图 2.6 所示。

图 2.6　程序执行结果

【分析讨论】

- 第 30、31 句创建了两个基于文本的按钮。其中,第一个显示字符串

"Run",第二个显示字符串"Exit"。
- 第33、35句分别设置了动作事件处理程序。
- 按钮响应 ActionEvent 事件,为注册这些事件的处理程序,对按钮调用了 setOnAction( )方法,它使用了向方法传递 lambda 表达式的方式。此时,该方法的参数类型将提供 lambda 表达式的目标上下文。在 handler( )方法内,设置了 response 标签的文本以反映 Run 按钮被单击。这里是通过对标签调用 setText( )方法实现的。Exit 按钮的事件处理方式与此相同。
- 设置事件处理程序后,调用 addAll( )方法将 response 标签以及 btnRun 和 btnExit 按钮添加到场景中。addAll( )方法将调用父节点添加节点列表。
- 第20句用于在窗口中显示控件的方式。出于画面美观的角度,FlowPane 构造方法传递了两个值,指定了场景中元素周围的水平和垂直间隙。第22句用于设置 FlowPane 中元素的堆砌方式,即元素居中对齐。Pos 是一个指定对齐常量的枚举类型,包含在 javafx.geometry 包中。

2.5 小结

JavaFx 提供了一个强大的、流线化且灵活的框架,简化了视觉效果出色的 GUI 的开发。JavaFX 的提出是为了取代 Swing,并定位于未来的开发平台。预计在未来几年中,JavaFX 将会逐渐取代 Swing 并应用到新的项目中,因此任何 Java 程序员都应该重视 JavaFX 的应用开发。

第 3 章 JavaFX 控件——Image、ImageView 与 TreeView

JavaFX 的控件中允许包含图片，还可以在场景中直接嵌入独立的图片。JavaFX 对图片支持的基础是 Image 和 ImageView 这两个类。Image 封装了图片，而 imageView 管理图片的显示。这两个类包含在 javafx.scene.image 包中。在 JavaFX 中，TreeView 以树状形式显示数据的分层视图。本章将介绍基于 NetBeans IDE 开发包含图片以及 TreeView 的 JavaFX 应用的方法。

3.1 Image 和 ImageView 控件

Image 类可以从 InputStream、URL 或图片文件的路径中加载图片。Image 类的构造方法为 Image(String url)。其中，url 指定 URL 或图片文件的路径；如果参数的格式不正确，则认为该参数指向一个路径，否则从 URL 位置加载图片。注意：Image 没有继承 Node，所以它不能作为场景图的一部分。ImageView 的构造方法为 ImageView(Image image)。

【例 3.1】 加载一幅清华大学出版社网站的图片，使用 ImageView 将该图片显示出来。

```
1.  package imagedemo;
2.  import javafx.application.Application;
3.  import javafx.scene.Scene;
4.  import javafx.scene.layout.StackPane;
5.  import javafx.stage.Stage;
6.  import javafx.scene.image.Image;
7.  import javafx.scene.image.ImageView;
8.  public class imageDemo extends Application {
9.      @Override
```

```
10.     public void start(Stage primaryStage) {
11.         //创建 Image 与 ImageView 对象
12.         Image image = new Image("http://www.tup.com.cn/upload/kindeditor/image/20220505/lidh2022 0505134151_8151.jpg");
13.         ImageView imageView = new ImageView();
14.         imageView.setImage(image);
15.         //显示图片
16.         StackPane root = new StackPane();
17.         root.getChildren().add(imageView);
18.         Scene scene = new Scene(root, 300, 250);
19.         primaryStage.setTitle("Image Read Test");
20.         primaryStage.setScene(scene);
21.         primaryStage.show();
22.     }
23.     public static void main(String[] args) {
24.         launch(args);
25.     }
26. }
```

【执行结果】

执行结果如图 3.1 所示。

图 3.1　程序的执行结果

【分析讨论】

- 第 12 句创建了一个 Image。但是，图片是不能添加到场景中的，必须先嵌入一个 ImageView 中(第 13 句)。

3.2 TreeView 控件

在 JavaFX 中，TreeView 以树状形式显示数据的分层视图。这里，分层是指一些条目是其他条目的子项。例如，在树用于显示文件系统的情形下，单独的文件从属于包含它们的目录。在 TreeView 中，用户可以根据需要展开或收缩树枝，这样就可以以一种紧凑但可以展开的形式显示分层数据。TreeView 实现了一种概念上的基于树的数据结构。树从根节点开始，根节点指出树的起点。在根节点下有一个或多个子节点。子节点分为叶子节点（终端节点不包含子节点）和树枝节点（构成子树的根节点，子树是包含在更大的树结构中的树）。从根节点到某个特定节点的节点序列称为路径。当树的大小超出视图的尺寸时，TreeView 将会自动提供滚动条。根据需要自动添加滚动条能够节省空间。

TreeView 是泛型类，其声明如下。

```
class TreeView<T>
```

- T——指定树中条目保存值的类型。一般为 String 类型。

TreeView 的构造方法定义如下。

```
TreeView(TreeItem<T> rootNode)
```

- rootNode——子树的根节点。因为所有的节点都派生自根节点，所以根节点是唯一需要传递给 TreeView 的节点。
- TreeItem——构成树的条目是 TreeItem 类型的对象。TreeItem 没有继承 Node，所以 treeItem 对象不是通用对象，它可以用在 TreeView 中，但不能作为独立控件使用。

TreeItem 类的声明为 class TreeItem<T>

- T——指定了 TreeItem 保存值的类型。

使用 TreeView 的方法如下。

① 构造要显示的树。首先，创建根节点；然后，向根节点添加其他节点，可以通过对 getChildren( )方法返回的列表调用 add( )或 addAll( )方法实现，添加的节点可以是叶子节点或子树。

② 构造完成树以后，将其根节点传递给 TreeView 的构造方法以创建 TreeView 对象。

③ 处理 TreeView 中选择的事件。首先，调用 getSelectinModel( )方法获

得选择模式，然后调用 selectItemProperty( )方法获得选中的属性，最后，通过对该方法的返回值调用 addListener( )方法以添加事件监听器。每次做出选择时，就将对新选项的引用作为新值传递给 changed( )方法。

④ 通过调用 getValue( )方法可以获得 treeItem 的值。还可以前向或后向沿着某个条目的树路径前进。

⑤ 通过调用 getParent( )方法可以得到某个父节点。调用 getChildren( )方法可以得到某个节点的子节点。

【例3.2】 创建一个树，显示一个食物层次。树中存储 String 类型的条目。根节点的标签是 Food。根节点有 3 个直接子节点——水果、蔬菜和坚果。水果节点包含 3 个子节点——苹果、梨和橘子。苹果节点下有 3 个叶子节点——富士、国光和红玉。每次做出选择时都显示所选项的名称。

```
1.  package treeviewdemo;
2.  import javafx.application.*;
3.  import javafx.scene.*;
4.  import javafx.stage.*;
5.  import javafx.scene.layout.*;
6.  import javafx.scene.control.*;
7.  import javafx.event.*;
8.  import javafx.beans.value.*;
9.  import javafx.geometry.*;
10. public class TreeViewDemo extends Application {
11. Label response;
12. public static void main(String[ ] args) {
13.     //通过调用 launch()方法启动 JavaFX 应用
14.     launch(args);
15. }
16. //重写 start()方法
17. public void start(Stage myStage) {
18.     //设置舞台的标题
19.     myStage.setTitle("Demonstrate a TreeView");
20.     //使用 FlowPane 布局
21.     //指定场景中元素周围的水平和垂直间隙
22.     FlowPane rootNode = new FlowPane(10, 10);
23.     //中心对齐
24.     rootNode.setAlignment(Pos.CENTER);
25.     //创建一个场景
```

```
26.     Scene myScene = new Scene(rootNode, 310, 460);
27.     //在舞台中设置场景
28.     myStage.setScene(myScene);
29.     //创建一个标签提示用户的选择项
30.     response = new Label("No Selection");
31.     //创建树的根节点
32.     TreeItem<String> tiRoot = new TreeItem<String>("Food");
33.     //创建水果子节点
34.     TreeItem<String> tiFruit = new TreeItem<String>("水果");
35.     //构造苹果子树
36.     TreeItem<String> tiApples = new TreeItem<String>("苹果");
37.     //将不同品种的苹果添加到苹果子树节点
38.     tiApples.getChildren().add(new TreeItem<String>("富士"));
39.     tiApples.getChildren().add(new TreeItem<String>("国光"));
40.     tiApples.getChildren().add(new TreeItem<String>("红玉"));
41.     //将不同的水果添加到水果子树节点 Add varieties to the fruit node
42.     tiFruit.getChildren().add(tiApples);
43.     tiFruit.getChildren().add(new TreeItem<String>("梨"));
44.     tiFruit.getChildren().add(new TreeItem<String>("橘子"));
45.     //最后,将水果子节点添加到根节点
46.     tiRoot.getChildren().add(tiFruit);
47.     //现在,用同样的方法构造蔬菜子树
48.     TreeItem<String> tiVegetables = new TreeItem<String>("蔬菜");
49.     tiVegetables.getChildren().add(new TreeItem<String>("玉米"));
50.     tiVegetables.getChildren().add(new TreeItem<String>("豌豆"));
51.     tiVegetables.getChildren().add(new TreeItem<String>("西兰花"));
52.     tiVegetables.getChildren().add(new TreeItem<String>("豆颈"));
53.     tiRoot.getChildren().add(tiVegetables);
54.     //构造坚果子树节点
55.     TreeItem<String> tiNuts = new TreeItem<String>("坚果");
56.     tiNuts.getChildren().add(new TreeItem<String>("核头"));
57.     tiNuts.getChildren().add(new TreeItem<String>("花生"));
58.     tiNuts.getChildren().add(new TreeItem<String>("山核头"));
59.     tiRoot.getChildren().add(tiNuts);
60.     //用创建的树创建 TreeView
61.     TreeView<String> tvFood = new TreeView<String>(tiRoot);
62.     //设置 TreeView 的选择模式
63.      MultipleSelectionModel< TreeItem< String > > tvSelModel =
    tvFood.getSelectionModel();
```

```
64.        //用变化监听器响应用户选择的一条 TreeView
65.        tvSelModel.selectedItemProperty().addListener(new
           ChangeListener<TreeItem<String>>() {
66.            public void changed(ObservableValue<? extends TreeItem<
           String>> changed,
67.                                TreeItem<String> oldVal, TreeItem<
           String> newVal) {
68.            //显示用户的选择以及子树路径
69.            if(newVal != null) {
70.              //构造入口路径与选择的条目
71.              String path = newVal.getValue();
72.              TreeItem<String> tmp = newVal.getParent();
73.              while(tmp != null) {
74.                path = tmp.getValue() + " -> " + path;
75.                tmp = tmp.getParent();
76.              }
77.              //显示用户选择的条目以及路径
78.              response.setText("Selection is " + newVal.getValue()
           + "\nComplete path is " + path);
79.            }
80.        }});
81.        //将树根节点添加到场景中
82.        rootNode.getChildren().addAll(tvFood, response);
83.        //在舞台中显示场景
84.        myStage.show();
85.        }
86. }
```

【执行结果】

执行结果如图 3.2 所示。

【分析讨论】

- 第 32 句创建了树的根节点；其次，创建了根节点之下的节点，这些节点构成了子树的根节点。一个表示水果（第 34 句），一个表示蔬菜（第 48 句），一个表示坚果（第 55 句）。然后，为这些子树添加叶子节点。其中，水果子树还包含一个子树，它包含不同品牌的苹果（第 38～40 句）。这里关键的知识点是树中的每个树枝要么走向一个叶子节点，要么走向一个子树的根节点。

- 构造所有的节点之后，通过对根节点调用 add() 方法，就可以将每个子

第 3 章　JavaFX 控件——Image、ImageView 与 TreeView

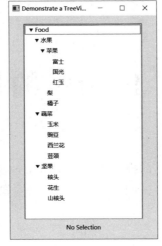

图 3.2　程序的执行结果

　　树的根节点添加到树的根节点(第 38～40 句,第 42～44 句,第 49～53 句,第 56～59 句)。
- 在事件处理监听程序中,从根节点到选定节点的路径通过第 71～75 句实现。首先,获取选中节点的值(一个字符串,即节点的名称)。然后,创建一个 TreeItem<String>类型的变量,并将其初始化为引用新选中节点的父节点。如果新选中的节点没有父节点,那么其值为 NULL。否则,进入循环,将每个父节点的值添加到 path 中。这个过程不断循环进行,直到找到树的根节点。

3.3　小结

　　JavaFX 对图片支持的基础是 Image 和 ImageView 这两个类。Image 封装了图片,而 imageView 则负责管理图片的显示。在 JavaFX 中,TreeView 以树状形式显示数据的分层视图。在 TreeView 中,用户可以根据需要展开或收缩树枝,这样就可以用一种紧凑但可以展开的形式显示分层数据。TreeView 实现了一种概念上的基于树的数据结构。本章介绍了基于 NetBeans IDE 开发包含图片以及 TreeView 的 JavaFX 应用程序的方法。希望读者认真领会理解,通过实际的上机调试掌握这些示例程序的编程方法。

Chapter 4
第4章 JavaFX 的其他控件

JavaFX 定义了一组丰富的控件,它们包含在 javafx.scene.control 包中。本节将介绍复选框(CheckBox)、列表(ListView)和文本框(TextField)这 3 个控件的用法。

4.1 CheckBox

在 JavaFX 中,CheckBox 类封装了复选框的功能,它的父类是 ButtonBase。复选框控件支持 3 种状态——选中、未选中以及不确定(indeterminate)(也称未定义(undefined),用于表示复选框的状态尚未被设置,或者对于特定的情形不重要)。CheckBox 的构造方法如下。

CheckBox(String str)——用 str 指定的文本作为标签。当勾选复选框时将会产生动作事件。

【例 4.1】 复选框的实例。

```
1.  package CheckBoxDemo;
2.  import javafx.application.*;
3.  import javafx.scene.*;
4.  import javafx.stage.*;
5.  import javafx.scene.layout.*;
6.  import javafx.scene.control.*;
7.  import javafx.event.*;
8.  import javafx.geometry.*;
9.  public class CheckBoxDemo extends Application {
10.     CheckBox cbSmartphone;
11.     CheckBox cbTablet;
12.     CheckBox cbNotebook;
13.     CheckBox cbDesktop;
14.     Label response;
```

```
15.    Label selected;
16.    String computers;
17.    public static void main(String[ ] args) {
18.      //通过调用 launch()方法启动 JavaFX 应用
19.      launch(args);
20.    }
21.    //重写 start()方法
22.    public void start(Stage myStage) {
23.      //设定舞台标题
24.      myStage.setTitle("Demonstrate Check Boxes");
25.      //设定根节点 FlowPane 布局
26.      FlowPane rootNode = new FlowPane(Orientation.VERTICAL, 10, 10);
27.      //指定场景中的元素居中对齐
28.      rootNode.setAlignment(Pos.CENTER);
29.      //创建场景
30.      Scene myScene = new Scene(rootNode, 230, 200);
31.      //在舞台中设定场景
32.      myStage.setScene(myScene);
33.      Label heading = new Label("你喜欢哪一种移动手机?");
34.      //创建一个标签,以报告复选框的变化
35.      response = new Label("");
36.      //创建一个标签以报告所有被选中的复选框
37.      selected = new Label("");
38.      //创建复选框对象
39.      cbSmartphone = new CheckBox("华为");
40.      cbTablet = new CheckBox("小米");
41.      cbNotebook = new CheckBox("中兴");
42.      cbDesktop = new CheckBox("联想");
43.      //处理复选框事件
44.      cbSmartphone.setOnAction(new EventHandler<ActionEvent>() {
45.        public void handle(ActionEvent ae) {
46.          if(cbSmartphone.isSelected())
47.            response.setText("华为 was just selected.");
48.          else
49.            response.setText("华为 was just cleared.");
50.          showAll();
51.        }
52.      });
```

```
53.    cbTablet.setOnAction(new EventHandler<ActionEvent>() {
54.       public void handle(ActionEvent ae) {
55.         if(cbTablet.isSelected())
56.            response.setText("小米 was just selected.");
57.         else
58.            response.setText("小米 was just cleared.");
59.         showAll();
60.       }
61.    });
62.    cbNotebook.setOnAction(new EventHandler<ActionEvent>() {
63.       public void handle(ActionEvent ae) {
64.         if(cbNotebook.isSelected())
65.            response.setText("中兴 was just selected.");
66.         else
67.            response.setText("中兴 was just cleared.");
68.         showAll();
69.       }
70.    });
71.    cbDesktop.setOnAction(new EventHandler<ActionEvent>() {
72.       public void handle(ActionEvent ae) {
73.         if(cbDesktop.isSelected())
74.            response.setText("联想 was just selected.");
75.         else
76.            response.setText("联想 was just cleared.");
77.         showAll();
78.       }
79.    });
80.    //Add controls to the scene graph.
81.    rootNode.getChildren().addAll(heading, cbSmartphone, cbTablet,
82.                    cbNotebook, cbDesktop, response, selected);
83.    //Show the stage and its scene.
84.    myStage.show();
85.    showAll();
86.  }
87.  //Update and show the selections.
88.  void showAll() {
89.    computers = "";
```

```
90.     if(cbSmartphone.isSelected()) computers = "华为 ";
91.     if(cbTablet.isSelected()) computers += "小米 ";
92.     if(cbNotebook.isSelected()) computers += "中兴 ";
93.     if(cbDesktop.isSelected()) computers += "联想";
94.     selected.setText("移动电话 selected: " + computers);
95.   }
96. }
```

【执行结果】

执行结果如图 4.1 所示。

图 4.1　JavaFX 程序执行结果

【分析讨论】

- 该程序演示了复选框的使用。程序显示了 4 个复选框，表示手机的几种不同品牌。每次勾选复选框时都将产生 ActionEvent，这些事件的处理程序首先报告复选框是被勾选还是被清除。为此，它们对事件源调用 isSelected( ) 方法。如果被勾选，则返回 true；如果被清除，则返回 false。接下来，程序调用了 showAll( ) 方法，显示所有已经被勾选的复选框。
- 默认情形下，FlowPane 的布局是水平流式。程序中通过 Orientation.VERTICAL 值作为第一个实际参数传递给 FlowPane 的构造方法以创建垂直的流式布局。

4.2　ListView

在 JavaFX 中，ListView 由 ListView 类封装。ListView 类可以显示一个选项列表，用户可以从中选择一个或多个选项。当列表中选项的数量超出控件空间可以显示的数量时，将会自动添加滚动条。ListView 是一个泛型类，其声明如下：

```
class ListView<T>
```

- T 用于指定列表视图中存储的选项的类型。
- 一般地,ListView 的构造方法的定义如下。

```
ListView(ObservableList<T> list)
```

List 指定了将要显示的选项列表,是一个 ObservableList 类型的对象。默认地,ListView 只允许在列表中一次选择一项。

创建在 ListView 中使用的 ObservableList 可以使用 FXCollection 类(包含在 javafx.collections 包中)定义的静态方法 observableArrayList( )。

```
static <E> ObservableList<E> observableArrayList(E ... elements)
```

E——指定了元素类型,元素通过 elements 传递。
如果希望自身设置选择的高度和宽度,可以调用以下两个方法:

```
final void setPrefHeight(double height)
final void setPrefwidth(double width)
```

还有一种同时设置高度和宽度的方法:

```
void setPrefSize(double width, double height)
```

使用 ListView 有两种基本方法。首先,可以忽略列表产生的事件,而是在程序需要时获得列表中的选中项;其次,通过注册事件监听器监视列表中的变化,每当用户改变列表中的选中项时,就可以做出响应。

事件监听器包含在 javafx.beans.value 包的 ChangeListener 接口中,该接口定义了 changed( )方法:

```
void changed (ObservableValue <? extends T > changed, T oldval, T newVal)
```

- changed——ObservableValue<T>的实例,而 ObservableValue<T>封装了可以观察的对象。
- oldValue 和 newValue——分别传递前一个值和新值。newValue 保存的是已被选中的列表选项的引用。

为了监听变化的事件,必须获得 ListView 使用模式,这是通过调用

getSelectionModel( )方法实现的。

```
final MultipleSelectionModel<T> getSelectionModel()
```

该方法返回对模式的引用。

该方法继承了 SelectionModel 类,定义了多项选择使用的模式。

只有打开多项选择模式后,ListView 才允许进行多项选择。

使用 getSelectionModel( )方法返回的模式将获得对选中项属性的引用,该属性定义了选中列表中的元素将要发生什么。这是通过 selectedItemProperty( )方法实现的。

```
final ReadOnlyObjectProperty<T> selectedItemProperty()
```

对返回的属性调用 addListener( )方法,将事件监听器添加给这个属性。

```
void addListener(ChangeListener<? super T> listener)
```

T——指定属性的类型。

【例 4.2】 创建一个显示多种计算机类型的列表视图,允许用户从中做出选择。用户选择一种类型后,就显示所选的项。

```
1.  package ListViewDemo;
2.  import javafx.application.*;
3.  import javafx.scene.*;
4.  import javafx.stage.*;
5.  import javafx.scene.layout.*;
6.  import javafx.scene.control.*;
7.  import javafx.geometry.*;
8.  import javafx.beans.value.*;
9.  import javafx.collections.*;
10. public class ListViewDemo extends Application {
11.     Label response;
12.     public static void main(String[ ] args) {
13.         //通过调用 launch()方法启动 JavaFX 应用
14.         launch(args);
15.     }
16.     //重写 start()方法
17.     public void start(Stage myStage) {
18.         //设置舞台的标题
```

```
19.        myStage.setTitle("ListView Demo");
20.        //对于 root node 使用 FlowPane 布局
21.        //指定场景中元素周围的水平和垂直间隙
22.        FlowPane rootNode = new FlowPane(10, 10);
23.        //指定元素居中对齐
24.        rootNode.setAlignment(Pos.CENTER);
25.        //创建一个场景
26.        Scene myScene = new Scene(rootNode, 200, 120);
27.        //在舞台中设置场景
28.        myStage.setScene(myScene);
29.        //创建一个标签
30.        response = new Label("Select Computer Type");
31.        //创建一个字符串列表,并用 ObservableList 初始化 ListView
32.          ObservableList< String > computerTypes = FXCollections.observableArrayList ( " Smartphone ", " Tablet ", " Notebook ", " Desktop" );
33.        //Create the list view.
34.          ListView< String > lvComputers = new ListView< String >(computerTypes);
35.        //设置 lvComputers 控件的首选宽度和高度
36.        lvComputers.setPrefSize(100, 70);
37.        //获得 lvComputers 控件的选择模式
38.        MultipleSelectionModel<String> lvSelModel = lvComputers.getSelectionModel();
39.          //ListView 使用 MultipleSelectionModel 模式, 调用 selectedItemProperty()方法注册变化监听器
40.          lvSelModel.selectedItemProperty ( ).addListener ( new ChangeListener<String>() {
41.          public void changed(ObservableValue<? extends String> changed, String oldVal, String newVal) {
42.        //显示被选择的项
43.          response.setText("Computer selected is " + newVal);
44.        }
45.        });
46.        //在场景中添加标签与 ListView
47.        rootNode.getChildren().addAll(lvComputers, response);
48.        //在舞台中显示这个场景
49.        myStage.show();
50.        }
51.    }
```

【执行结果】

执行结果如图 4.2 所示。

【分析讨论】

- 当 ListView 中的内容超过控件大小时，就会自动添加一个滚动条。
- 第 32 句创建了一个字符串列表，并用 ObservableList 初始化 ListView。之后，第 36 句设置了控件的宽度和高度。
- 第 34 句创建了一个 ListView 对象，第 36 句设置了这个 ListView 控件的宽度与高度。
- 第 38 句获得了 lvComputers 控件的选择模式。第 40 句的 ListView 使用了 MultipleSelectionModel，并模式调用 selectedItemProperty( )方法注册变化监听器。第 43 句显示了被选择的项。第 47 句在场景中添加了标签与 ListView。第 49 句在舞台中显示了这个场景。

图 4.2 JavaFX 程序执行结果

4.3 TextField

当用户需要输入字符串时，JavaFX 提供了 TextField 控件，用于输入一行文本，例如获得名称、ID 字符串、地址等。TextField 继承了 TextInputControl。TextField 定义了两个构造方法。第一个是默认的构造方法，用于创建一个具有默认大小的空文本框。第二个构造方法可以指定文本框的初始内容。当需要指定文本框的大小时，可以通过调用下列方法实现：

final void setColumnCount(int columns)——Columns 的值用来确定 textField 的大小。

setText( )方法可以设置文本框中的文本，getText( )方法可以获取当前文本。当用户需要在文本框中显示一条提示消息时，可以调用如下方法：

final voidsetPromptText(String str)——str 是在文本框中显示的提示信息，这个字符串将用低颜色强度（灰色色调）显示。

【例 4.3】 创建一个需要输入搜索字符串的文本框，当用户在文本框中具有输入焦点时，按 Enter 键或者单击 Get Name 按钮就会获取并显示该字符串。

```
1.  package textfielddemo;
2.  import javafx.application.*;
3.  import javafx.scene.*;
```

```
4.   import javafx.stage.*;
5.   import javafx.scene.layout.*;
6.   import javafx.scene.control.*;
7.   import javafx.event.*;
8.   import javafx.geometry.*;
9.   public class TextFieldDemo extends Application {
10.     TextField tf;
11.     Label response;
12.     public static void main(String[ ] args) {
13.       //通过调用 launch()方法启动 JavaFX 应用
14.       launch(args);
15.     }
16.     //重写 start()方法
17.     public void start(Stage myStage) {
18.       //设置舞台标题
19.       myStage.setTitle("Demonstrate a TextField");
20.       //使用 FlowPane 布局
21.       //设置场景中元素周围的水平和垂直间隙
22.       FlowPane rootNode = new FlowPane(10, 10);
23.       //Center the controls in the scene.
24.       rootNode.setAlignment(Pos.CENTER);
25.       //创建一个场景
26.       Scene myScene = new Scene(rootNode, 230, 140);
27.       //在舞台中设置场景
28.       myStage.setScene(myScene);
29.       //Create a label that will report the state of the selected check box.
30.       response = new Label("Enter Name: ");
31.       //创建一个按钮
32.       Button btnGetText = new Button("Get Name");
33.       //创建文本框
34.       tf = new TextField();
35.       //设置文本框的提示信息
36.       tf.setPromptText("Enter a name.");
37.       //设置文本框的宽度
38.       tf.setPrefColumnCount(15);
39.       //使用 Lambda 表达式处理文本框的动作事件
40.       tf.setOnAction( (ae) -> response.setText("Enter pressed. Name is: " + tf.getText()));
```

```
41.        //当单击按钮时,使用 lambda 表达式得到文本框中的文本
42.        btnGetText.setOnAction((ae) -> response.setText("Button
   pressed. Name is: " + tf.getText()));
43.        //Use a separator to better organize the layout.
44.        Separator separator = new Separator();
45.        separator.setPrefWidth(180);
46.        //Add controls to the scene graph.
47.        rootNode.getChildren().addAll(tf, btnGetText, separator,
   response);
48.        //Show the stage and its scene.
49.        myStage.show();
50.      }
51. }
```

【执行结果】

执行结果如图 4.3 所示。

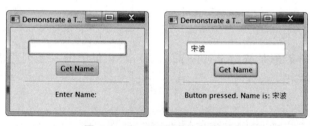

图 4.3　JavaFX 程序执行结果

【分析讨论】

- 本示例将 Lambda 表达式作为事件处理程序。每个处理程序均由单个方法调用组成,这就使得事件处理程序成为 Lambda 表达式的完美实现。

4.4　小结

JavaFX 定义了一组丰富的控件,它们包含在 javafx.scene.control 包中。本节介绍了复选框(CheckBox)、列表(ListView)和文本框(TextField)这 3 个控件的用法。读者可以通过实际的上机操作理解与掌握这种程序的编程方法。

Chapter 5 第 5 章 JavaFX 菜单

菜单是 GUI 的重要组成部分,它可以让用户访问程序的核心功能,所以 JavaFX 为菜单提供了广泛的支持。JavaFX 的一个主要优势在于通过使用效果/变换改变控件的精确外观。通过这个功能,可以让 GUI 具有用户期望的复杂现代外观。本章将介绍开发 JavaFX 菜单应用程序以及让 GUI 具有用户期望的外观的原理与方法。

5.1 基础知识

JavaFX 的菜单系统由 javafx.scene.control 包中的一系列相关的类提供支持,如表 5.1 所示。

表 5.1 JavaFX 的核心菜单类

类	主要功能
CheckMenuItem	复选菜单项
ContextMenu	弹出菜单
Menu	标准菜单,由一个或多个 MenuItem 组成
MenuBar	保存程序的顶级菜单的对象
MenuItem	填充菜单的对象
RadioMenuItem	单选菜单项
SeparatorMenuItem	菜单项之间的可视分隔符

- 如果要创建程序的顶级菜单,首先要创建一个 MenuBar 实例,即这个类是菜单的容器。在 MenuBar 实例中,将添加 Menu 实例。每个 Menu 对象定义了一个菜单,也就是说,每个 Menu 对象包含一个或多个可以选择的菜单项。Menu 显示的菜单项是 MenuItem 类型的对象。因此,

MenuItem 定义了用户可以选择的选项。
- 除了标准菜单项,还可以在菜单中包含复选菜单项和单选菜单项,它们的操作与复选框和单选按钮控件类似。复选菜单项用 CheckMenuItem 类创建,单选菜单项用 RadioMenuItem 类创建。这两个类扩展了 MenuItem 类。
- SeparatorMenuItem 类用于在菜单中创建一条分隔线,它继承了 CustomMenuItem 类,后者使得在菜单中嵌入其他类型的控件变得容易。CustomMenuItem 类扩展了 MenuItem 类。
- MenuItem 类没有继承 Node 类。因此,MenuItem 类的实例只能用在菜单中,而不能以其他方式加入场景图中。但是,MenuBar 类继承了 Node 类,所以可以把菜单栏添加到场景图中。MenuItem 是 Menu 的超类,它可以创建子菜单,也就是菜单中的菜单。要创建子菜单,首先要创建一个 Menu 对象,并用 MenuItem 填充它,然后把它添加到另一个 Menu 对象中。
- 选择菜单项后,会生成动作事件。与所选项关联的文本称为选择的名称,所以不需要通过检查名称确定哪个菜单项被选择。
- 也可以创建独立的上下文菜单,它们在激活时会被弹出。首先,需要创建一个 ContextMenu 类的对象。然后,向该对象添加 MenuItem。如果为某个控件定义了上下文菜单,那么激活该菜单的方式通常是在该控件上右击。ContextMenu 类继承了 PopupControl 类。
- 工具栏是与菜单相关的一种特性。工具栏由 ToolBar 类支持。该类创建独立的组件,通常用于快速访问程序菜单中包含的功能。

5.2 MenuBar、Menu 和 MenuItem 概述

要为程序创建菜单,最少要用到 MenuBar、Menu 以及 MenuItem 这 3 个类。上下文菜单也会用到 MenuItem。因此,这 3 个类是菜单系统的基础。

1. MenuBar

Menubar 是菜单的容器,它是为程序提供主菜单的控件。MenuBar 类继承了 Node 类,因此可以把它添加到场景图中。MenuBar 有两个构造方法,第一个是默认的构造方法,需要在使用之前在其中填充菜单;第二个构造方法允许指定初始的菜单栏列表。一般地,程序有且只有一个菜单栏。Menubar 定义的方法中,getMenus( )方法经常被使用,它返回一个由菜单栏管理的菜单列表。创建的菜单将被添加到这个列表中。

```
final ObservableList<Menu> getMenus( )
```

调用 add( )方法可以把 Menu 实例添加到这个菜单列表中,也可以用 addAll( )方法在一次调用中可以添加两个或多个 Menu 实例。添加的菜单将按照添加顺序从左到右排列在菜单中。如果要在特定位置添加一个菜单,则可以使用以下 add( )方法。

```
void add(int idx, Menu menu)
```

- Menu 将被添加到由 idx 指定的索引位置。索引从 0 开始,0 对应最左边的菜单。

当需要删除不再需要的菜单时,可以通过对 getMenus( )方法返回的 ObservableList 调用 remove( )方法实现。该方法的两种定义形式如下:

```
void remove(Menu menu)
void remove(int idx)
```

- Menu——对要删除的菜单的引用,idx 是要删除的菜单的索引,索引从 0 开始。如果找到并删除了菜单项,则第一种形式返回 true,第二种形式返回对删除元素的引用。

2. Menu

Menu 封装了菜单,菜单项用 MenyItem 填充。而 Menu 派生自 MenuItem,这意味着一个 Menu 实例可以是另一个 Menu 实例中的选项,从而能够创建菜单的子菜单。Menu 定义了以下 4 个构造方法。

- Menu(String name)——该构造方法创建的菜单具有 name 指定的名称。
- Menu(String name, Node image)——image 指定了要显示的图片。
- Menu(String name, Node inage, MenuItem... menuItems)——允许指定最初的添加菜单项列表。
- Menu( )——可以用默认的构造方法创建未命名的菜单。然后,创建菜单后再调用 setText( )方法添加名称,调用 setGraphic( )方法添加图片。

每个菜单都维护一个由它包含的菜单项组成的列表。要在菜单中添加菜单项,需要把菜单添加到这个列表中。可以在 Menu 的构造方法中指定它们,或者把它们添加到列表中。为此,首先调用 getItem( )方法:

```
final ObservableList<MenuItem> getItems( )
```

该方法返回当前与菜单相关联的菜单项列表。然后,调用 add( )或 addAll( )方法把菜单项添加到这个列表中。另外,也可以调用 remove( )方法从中删除菜单项,调用 size( )方法获取列表的大小。此外,可以在菜单项列表中添加一条菜单分隔线,该分隔线是 SwparatorMenuItem 类型的对象。分隔线允许相关的菜单项分组,从而有助于组织菜单。分隔线可以帮助突出显示重要的菜单项。

3. MenuItem

MenuItem 封装了菜单中的元素,该元素可以是链接到某个程序动作的选项,也可以用于显示子菜单。MenuItem 定义了以下 3 种构造方法。

- MenuItem( )——创建一个空菜单项。
- MenuItem(String name)——用指定的名称创建菜单项。
- MenuItem(String name,Node image)——用包含的图片创建菜单项。

MenuItem 被选中时,将产生动作事件。通过调用 setOnAction( )方法可以为这种事件注册事件处理程序。

```
final void srtOnAction(EventHandler<ActionEvent> handle)
```

MenuItem 提供的 setDisable( )方法可以用来启用或禁用菜单项。

```
final void setDisable(boolean disable)
```

- 如果 disable 为 true,则禁用菜单项;如果 disable 为 false,则启用菜单项。

5.3 创建主菜单

一般地,主菜单是由菜单栏定义的菜单,也是定义了程序的全部功能的菜单。创建主菜单需要以下几个步骤:

① 创建用于保存菜单的 MenuBar 对象实例;

② 构造将包含在菜单栏中的每个菜单,首先创建一个 Menu 对象,然后向该对象添加 MneuItem;

③ 把菜单栏添加到场景图中;

④ 对于每个菜单项,添加动作事件处理程序,以响应选中菜单项时生成的动作事件。

【例5.1】 创建一个菜单栏,其中包含3个菜单。第一个是标准的 File 菜单,它包含 Open、Close、Save 和 Exit 选项;第二个是 Options 菜单,它包含 Colors 和 Priority 两个子菜单;第三个菜单称为 Help,它只需要 About 一个选项。选中一个菜单项时,将在一个标签中显示所选项的名称。

```java
1.   import javafx.application.*;
2.   import javafx.scene.*;
3.   import javafx.stage.*;
4.   import javafx.scene.layout.*;
5.   import javafx.scene.control.*;
6.   import javafx.event.*;
7.   import javafx.geometry.*;
8.   public class MenuDemo extends Application {
9.     Label response;
10.    public static void main(String[] args) {
11.      //通过调用 launch()方法启动 JavaFX 应用
12.      launch(args);
13.    }
14.    //重写 start()方法
15.    public void start(Stage myStage) {
16.      //设置舞台的标题
17.      myStage.setTitle("Demonstrate Menus");
18.      //定义根节点
19.      BorderPane rootNode = new BorderPane();
20.      //创建一个场景
21.      Scene myScene = new Scene(rootNode, 300, 300);
22.      //在舞台中设置场景
23.      myStage.setScene(myScene);
24.      //定义一个标签响应用户的选择
25.      response = new Label("Menu Demo");
26.      //创建 MenuBar 对象
27.      MenuBar mb = new MenuBar();
28.      //创建 File 菜单
29.      Menu fileMenu = new Menu("File");
30.      MenuItem open = new MenuItem("Open");
31.      MenuItem close = new MenuItem("Close");
32.      MenuItem save = new MenuItem("Save");
33.      MenuItem exit = new MenuItem("Exit");
```

```
34.    fileMenu.getItems().addAll(open, close, save, new
       SeparatorMenuItem(), exit);
35.    //将 File 菜单添加到 MenuBar 中
36.    mb.getMenus().add(fileMenu);
37.    //创建 Options 菜单
38.    Menu optionsMenu = new Menu("Options");
39.    //创建 Colors 子菜单
40.    Menu colorsMenu = new Menu("Colors");
41.    MenuItem red = new MenuItem("Red");
42.    MenuItem green = new MenuItem("Green");
43.    MenuItem blue = new MenuItem("Blue");
44.    colorsMenu.getItems().addAll(red, green, blue);
45.    optionsMenu.getItems().add(colorsMenu);
46.    //创建 Priority 子菜单
47.    Menu priorityMenu = new Menu("Priority");
48.    MenuItem high = new MenuItem("High");
49.    MenuItem low = new MenuItem("Low");
50.    priorityMenu.getItems().addAll(high, low);
51.    optionsMenu.getItems().add(priorityMenu);
52.    //添加分隔符
53.    optionsMenu.getItems().add(new SeparatorMenuItem());
54.    //创建 Reset 菜单项
55.    MenuItem reset = new MenuItem("Reset");
56.    optionsMenu.getItems().add(reset);
57.    //将 Options 菜单添加到 MenuBar
58.    mb.getMenus().add(optionsMenu);
59.    //创建 Help 菜单
60.    Menu helpMenu = new Menu("Help");
61.    MenuItem about = new MenuItem("About");
62.    helpMenu.getItems().add(about);
63.    //将 Help 菜单添加到 MenuBar 中
64.    mb.getMenus().add(helpMenu);
65.    //定义动作事件处理程序,以响应选中菜单项时生成的动作事件
66.    EventHandler<ActionEvent> MEHandler = new EventHandler<
       ActionEvent>() {
67.        public void handle(ActionEvent ae) {
68.            String name = ((MenuItem)ae.getTarget()).getText();
69.            //如果选择 Exit,则退出程序
```

```
70.         if(name.equals("Exit"))  Platform.exit();
71.         response.setText( name + " selected");
72.      }
73.   };
74.   //针对每个菜单项注册动作事件处理程序
75.   open.setOnAction(MEHandler);
76.   close.setOnAction(MEHandler);
77.   save.setOnAction(MEHandler);
78.   exit.setOnAction(MEHandler);
79.   red.setOnAction(MEHandler);
80.   green.setOnAction(MEHandler);
81.   blue.setOnAction(MEHandler);
82.   high.setOnAction(MEHandler);
83.   low.setOnAction(MEHandler);
84.   reset.setOnAction(MEHandler);
85.   about.setOnAction(MEHandler);
86.   //将 MenuBar 添加到窗口顶部
87.   //响应用户选择的标签显示在窗口中间
88.   rootNode.setTop(mb);
89.   rootNode.setCenter(response);
90.   //在窗口中显示舞台及其场景
91.   myStage.show();
92.   }
93. }
```

【执行结果】

执行结果如图 5.1 所示。

图 5.1　程序的执行结果

【分析讨论】

- 第 19 句创建的根节点的对象类型是 BorderPane,它定义了一个包含 5 个区域的窗口,这 5 个区域分别是顶部、底部、左侧、右侧和中央。
- 第 27 句用来构造菜单栏,此时,菜单栏是空的。第 29~33 句创建 File 菜单及其菜单项。第 34 句将各个菜单项添加到 File 菜单中。第 36 句将 File 菜单添加到菜单栏中。此时,菜单栏中将包含 File 菜单,File 菜单将包含 4 个选项:Open、Close、Save 和 Exit。
- 第 38 句将创建 Options 菜单,它包含 Colors 和 Oriority 两个子菜单,还包含 Reset 菜单项。第 40~44 句构造子菜单,第 45 句将它们添加到 Options 菜单中。第 47 句创建 Priority 子菜单,第 48~49 句创建两个菜单项 High 和 Low。第 50 句将它们添加到 Priority 子菜单中。第 51 句将 Priority 子菜单添加到 Options 菜单中。第 53 句在各个菜单项之间设置分隔符。第 55 句创建 Reset 菜单项。第 56 句将其添加到 Options 子菜单中。第 58 句将 Options 菜单添加到 MenuBar。
- 第 60 句创建 Help 菜单。第 61 句创建菜单项 About。第 62 句将其添加到 Help 菜单中。第 64 句将 Help 菜单添加到 MenuBar 中。
- 第 66~73 句定义动作事件处理程序,以响应选中菜单项时生成的动作事件。在 handle( ) 方法中,通过调用 getTarget( ) 方法获得事件的目标。该方法的返回类型是 MenuItem,其名称通过调用 getText( ) 方法返回。然后,这个字符串被赋值给 name。如果 name 包含字符串"Exit",就调用 platform.exit( ) 方法终止程序,否则在 response 标签中显示获得的名称。
- 第 75~85 句将 MEHandler 注册为每个菜单项的动作事件处理程序。第 88 句将菜单栏添加到根节点。

5.4 效果与变换

1. 效果

效果由 javafx.scene.effect 包中的 Effect 类及其子类支持。使用效果可以自定义场景图中节点的外观,如表 5.2 所示。

表 5.2 JavaFX 内置的效果

类	主要功能
Bloom	增加节点中较亮部分的亮度
BoxBlur	让节点变得模糊

续表

类	主要功能
DropShadow	在节点后面显示阴影
Glow	生成发光效果
InnerShadow	在节点内显示阴影
Lighting	创建光源的阴影效果
Reflection	显示倒影

2. 变换

变换由 javafx.scene.transform 包中的抽象类 Transform 支持,它有 4 个子类,分别是 Rotate、Scale、Shear 和 Translate。在节点上可以执行多种变换,例如可以旋转并缩放节点。Note 类支持变换。

【例 5.2】 程序创建了 4 个按钮,分别为 Rotate、Scale、Glow 和 Shadow。每单击一个按钮,就对按钮应用对应的效果或变换。

```
1.  import javafx.application.*;
2.  import javafx.scene.*;
3.  import javafx.stage.*;
4.  import javafx.scene.layout.*;
5.  import javafx.scene.control.*;
6.  import javafx.event.*;
7.  import javafx.geometry.*;
8.  import javafx.scene.transform.*;
9.  import javafx.scene.effect.*;
10. import javafx.scene.paint.*;
11. public class EffectsAndTransformsDemo extends Application {
12.     double angle = 0.0;
13.     double glowVal = 0.0;
14.     boolean shadow = false;
15.     double scaleFactor = 1.0;
16.     //定义一个基本的效果
17.     Glow glow = new Glow(0.0);
18.     InnerShadow innerShadow = new InnerShadow(10.0, Color.RED);
19.     Rotate rotate = new Rotate();
20.     Scale scale = new Scale(scaleFactor, scaleFactor);
21.     //创建 4 个按钮
```

```
22.     Button btnRotate = new Button("Rotate");
23.     Button btnGlow = new Button("Glow");
24.     Button btnShadow = new Button("Shadow off");
25.     Button btnScale = new Button("Scale");
26.     public static void main(String[ ] args) {
27.        //通过调用 launch()方法启动 JavaFX 应用
28.        launch(args);
29.     }
30.     //重写 start()方法
31.     public void start(Stage myStage) {
32.        //设置舞台的标题
33.        myStage.setTitle("Effects and Transforms Demo");
34.        //使用 FlowPane 布局定义根节点,并指定场景中元素周围的水平和垂直
               //间隙
35.        FlowPane rootNode = new FlowPane(10, 10);
36.        //Center the controls in the scene.
37.        rootNode.setAlignment(Pos.CENTER);
38.        //创建一个场景
39.        Scene myScene = new Scene(rootNode, 300, 100);
40.        //在舞台中设置场景
41.        myStage.setScene(myScene);
42.        //设置发光效果
43.        btnGlow.setEffect(glow);
44.        //将 Rotate 按钮添加到变换列表中
45.        btnRotate.getTransforms().add(rotate);
46.        //将 Scale 按钮添加到变换列表中
47.        btnScale.getTransforms().add(scale);
48.        //处理 Rotate 按钮的动作响应事件
49.        btnRotate.setOnAction(new EventHandler<ActionEvent>() {
50.           public void handle(ActionEvent ae) {
51.              //每当按钮被单击,它将旋转 30°
52.              //指定旋转的中心点
53.              angle += 30.0;
54.              rotate.setAngle(angle);
55.              rotate.setPivotX(btnRotate.getWidth()/2);
56.              rotate.setPivotY(btnRotate.getHeight()/2);
57.           }
58.        });
```

```
59.    //定义 scale 按钮的动作响应处理程序
60.    btnScale.setOnAction(new EventHandler<ActionEvent>() {
61.      public void handle(ActionEvent ae) {
62.        //每当按钮被点击,它的大小将发生变换
63.        scaleFactor += 0.1;
64.        if(scaleFactor > 1.0)   scaleFactor = 0.4;
65.        scale.setX(scaleFactor);
66.        scale.setY(scaleFactor);
67.      }
68.    });
69.    //定义 Glow 按钮的动作事件响应程序
70.    btnGlow.setOnAction(new EventHandler<ActionEvent>() {
71.      public void handle(ActionEvent ae) {
72.        //每当按钮被单击,它的颜色将逐渐变浅
73.        glowVal += 0.1;
74.        if(glowVal > 1.0) glowVal = 0.0;
75.        //Set the new glow value.
76.        glow.setLevel(glowVal);
77.      }
78.    });
79.    //定义 Shadow 按钮的动作事件的相应处理程序
80.    btnShadow.setOnAction(new EventHandler<ActionEvent>() {
81.      public void handle(ActionEvent ae) {
82.        //每当按钮被单击,它的颜色将逐渐变深
83.        shadow = !shadow;
84.        if(shadow) {
85.          btnShadow.setEffect(innerShadow);
86.          btnShadow.setText("Shadow on");
87.        } else {
88.          btnShadow.setEffect(null);
89.          btnShadow.setText("Shadow off");
90.        }
91.      }
92.    });
93.    //将标签与按钮添加到场景图中
94.    rootNode.getChildren().addAll(btnRotate, btnScale, btnGlow, btnShadow);
95.    //显示舞台和场景
```

```
96.    myStage.show();
97.  }
98. }
```

【执行结果】

执行结果如图 5.2 所示。

图 5.2 程序的执行结果

【分析讨论】

- 第 17～18 句定义了一个基本的效果。其中，Glow 生成的效果是节点具有发光的外观，构造方法为：

    Glow(double glowLevel)——glowLevel 用于指定光的亮度，取值范围在 0.0～1.0。创建 Glow 实例后，可以调用 setLevel( )方法改变发光的级别。

    final setLevel(double glowlevel)——glowLevel 用于指定光的亮度，取值范围在 0.0～1.0。

- InnerShadow 生成的效果是节点内具有阴影，其构造方法如下：

    InnerShadow(double radius，Color shadowColor)——radius 用于指定节点内阴影的半径，即指定阴影的大小。Shadow 用于指定阴影的颜色。Color 类型是 JavaFX 类型 javafx.scene.paint.Color，该类型定义了 Color.GREEN、ColorRED 和 Color.BLUE 等多个常量。

- 第 19～20 句定义了两个基本的变换。第 22～25 句定义了 4 个按钮。第 45 句将 Rotate 按钮添加到变换列表中。要向节点添加变换，可以把变换添加到节点维护的变换列表中。通过调用 Node 类定义的 getTransform( )方法可以获得该变换列表，如下所示：

    final ObservableList＜Transform＞ getTransform( )——该方法返回对变换列表的引用。要添加变换，只要调用 add( )方法把它添加到这个列表中即可。调用 clear( )方法可以清除该列表。调用 remove( )方法可以从列表中删除特定的元素。

- 为了演示变换，这里使用了 Rotale 类和 Scale 类。Rotale 绕着指定的点旋转节点。Rotate 类的构造方法如下：
    Rotate(double angle，double x，double y)——angle 用于指定旋转的角度。选中的中心点称为轴点，由 x 和 y 指定。
- 在创建 Rotate 实例之后(第 19 句)才设置这 3 个值。设置这 3 个值用到了如下 3 个方法(第 54~56 句)：

```
final void setAngle(double angle)
final void setPivotX(double x)
final void setPivotY(double y)
```

其中，angle 用于指定旋转的角度，x 和 y 用于指定旋转中心点。
- Scale 根据缩放因子缩放节点。Scale 类的构造方法如下：
    Scale(double widthFactor，double heightFactor)——widthFactor 用于指定对节点宽度应用缩放因子，heightFactor 用于指定对节点高度应用缩放因子。
- 创建 Scale 实例之后(第 20 句)，可以使用如下两个方法改变这两个因子(第 63~66 句)：

```
final void setX(double widthFactor)
final void setY(double heightFactor)
```

widthFactor 用于指定对节点的宽度应用缩放因子，heightFactor 用于指定对节点的高度应用缩放因子。

5.5 小结

JavaFX 提供了一个强大的、流线化且灵活的框架，简化了视觉效果出色的 GUI 的开发。本章比较详细地分析与讨论了 JavaFX 为菜单提供的支持。JavaFX 的一个主要优势在于通过使用效果/变换改变控件的精确外观。通过这个功能可以让 GUI 具有用户期望的复杂的现代外观。本章详细介绍了开发 JavaFX 菜单应用程序，以及让 GUI 具有用户期望的外观的基本原理与方法。

第6章 JavaFX Media 应用开发

在 Internet 上,媒体内容的活跃度持续增长,使得视频和音频已经成为 Rich Internet 应用的重要组成部分。如果能够充分利用 JavaFX 的多媒体功能,就可以极大地拓宽传统媒体的使用范围。JavaFX 的多媒体功能可以通过 JavaFX API 实现,JavaFX 提供的 Media 包使得开发人员能够创建多媒体应用,并在桌面窗口或支持的平台的网页中提供媒体播放功能。本章将通过一个综合示例介绍 JavaFX Media 的应用开发。

6.1 JavaFX 支持的媒体编解码器

JavaFX 多媒体功能支持的操作系统与 Java 运行时环境(JRE)认证系统的配置页面中列出的相同。有兴趣的读者可以参阅以下两个链接的详细信息:

- https://docs.oracle.com/en/java/javase/17/install/overview-jdk-installation.html
- https://www.oracle.com/java/technologies/javase/products-doc-jdk17certconfig.html

JavaFX 支持以下媒体编解码器格式。

- 音频:MP3——未压缩 PCM 的 AIFF;未压缩 PCM 的 WAV;带有高级音频的 MPEG-4 多媒体容器;编码(AAC)音频。
- 视频:VP6 视频和 MP3 音频的 FLV;带有 H.264/AVC(高级视频编码)视频压缩的 MPEG-4 多媒体容器。

FLV 容器由 JavaFX SDK 支持的平台的媒体堆栈支持。以这种格式编码的单个电影可以在受支持的平台上无缝地工作。服务器端需要标准 FLV MIME 的设置才能启用媒体流。对于所有操作系统都支持的 MPEG-4 多媒体容器,JavaFX SDK 也同样支持。

在 mac OS X 和 Windows 7 平台上,播放功能能够正常使用而无需额外的软件。但是,Linux 操作系统和早于 Windows 7 的 Windows 版本则需要安装第三方软件包,如认证系统配置页面所述,该页面的链接来自http://www.oracle.com/technetwork/java/javase/downloads/。AAC 和 H.264/AVC 解码具有特定的平台相关限制,如上述发布说明中 Java SE 下载页面所述。

　　某些音频和视频压缩类型的解码依赖于特定于操作系统的媒体引擎。JavaFX 媒体框架不会尝试处理这些本机引擎支持的所有多媒体容器格式和媒体编码。相反,该框架试图在支持 JavaFX 的所有平台上提供同等且经过良好测试的功能。

　　JavaFX 媒体堆栈支持的功能如下:
- 带有 MP3 和 VP6 的 FLV 容器;
- MP3 音频;
- 带有 AAC 和/或 H.264 的 MPEG-4 容器;
- HTTP,文件协议;
- 渐进式下载;
- Seeking;
- Buffer Progress;
- 播放功能(播放、暂停、停止、音量、静音、平衡、均衡器)。

6.2　HTTP 实时流媒体支持

　　通过添加 HTTP 实时流媒体支持可以下载播放列表文件,并使用 JavaFX Media 播放视频或音频片段。媒体播放器现在可以根据播放列表文件中指定的网络条件切换到备用流。对于给定的流,有一个播放列表文件和一组片段,流被分解成多个片段。该流可以是 MP3 的原始流,也可以是包含多路 AAC 音频和 H.264 视频的 MPEG-TS。当流是静态文件时,可以按需要播放流,或者当流是现场直播时,可以实现实时播放。在这两种情况下,流可以调整其比特率,对于视频,其分辨率也可以调整。

6.3　创建 Media Player

　　JavaFX 媒体的概念基于以下实体。
- 媒体——一种媒体资源,包含有关媒体的信息,例如其来源、分辨率和元数据。
- MediaPlayer——提供播放媒体控件的关键组件。
- MediaView——一个支持动画、半透明和效果的节点对象,媒体功能的

第 6 章 JavaFX Media 应用开发

每个元素都可以通过 JavaFX API 获得。

图 6.1 显示了 javafx.scene.media 包中的类。这些类相互依赖，并结合使用以创建嵌入式媒体播放器。

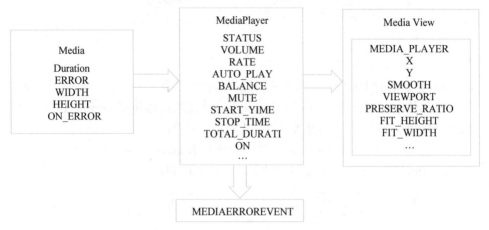

图 6.1　javafx.scene.media Package 中的类

MediaPlayer 类提供了控制媒体播放所需的所有属性和功能。用户可以设置自动播放模式，直接调用 PLAY() 方法或明确指定媒体应播放的次数。音量变量和平衡变量可分别用于调整音量水平和左右设置。音量范围为 0～1.0（最大值）。平衡范围从最左边的 −1.0、中间的 0 和右边的 1.0 连续。play()、stop() 和 pause() 方法控制媒体播放。此外，当用户执行以下操作之一时，一组方法将处理特定的事件。

- 缓冲数据。
- 到达媒体的尽头。
- 暂停，因为接收数据的速度不够快，所以无法继续播放。
- 遇到 MediaErrorEvent 类中定义的任何错误。

MediaView 类扩展了 Node 类，并提供了媒体播放器正在播放的媒体视图，它主要负责效果和转换，其 mediaPlayer 实例变量指定播放媒体的 mediaPlayer 对象。其他布尔属性用于应用节点类提供的特定效果，例如使媒体播放器能够旋转。

6.4　将媒体嵌入 Web Page

本节将探讨如何通过创建一个简单的媒体面板将动画媒体内容添加到 Web Page 中。要创建媒体播放器，需要实现图 6.2 所示的 3 个嵌套对象的结构。

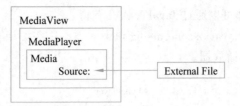

图 6.2　嵌套对象的结构

可以使用为创建 Java 应用程序而设计的任何开发工具构建 JavaFX 应用程序。本书使用的工具是 NetBeans IDE 13.0。在继续构建使用 JavaFX 媒体功能的示例应用程序之前,应当具备以下知识:

- 从 Java SE 下载页面下载并安装 JDK 9 和 NetBeans IDE 13.0;
- 参阅第 2 章的内容,能够创建 JavaFX 应用程序。

6.5　创建 JavaFX 应用

在 NetBeans IDE 中,按照以下方式创建与设置 JavaFX 项目。

(1) 在 File 菜单中选择 New Project 选项。

(2) 在 Categories 区域选择 Java With Ant,在 Projects 区域选择 JavaFX Application 选项,然后单击 Next 按钮。

(3) 将项目命名为 EmbeddedMediaPlayer,并确保创建应用程序类的字段值为 EmbeddedMediaPlayer——嵌入式媒体播放器,最后单击 Finish 按钮,如图 6.3 所示。

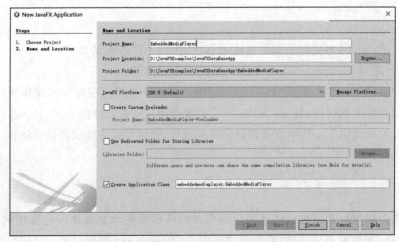

图 6.3　创建 EmbeddedMediaPlayer 项目

第 6 章　JavaFX Media 应用开发

（4）复制 Example 6-1 中的导入语句，并将它们粘贴到 EmbeddedMediaPlayer.java 文件中，替换 NetBeans IDE 自动生成的所有导入语句。

Example 6-1　Replace Default Import Statements

```java
import javafx.application.Application;
import javafx.scene.Group;
import javafx.scene.Scene;
import javafx.scene.media.Media;
import javafx.scene.media.MediaPlayer;
import javafx.scene.media.MediaView;
import javafx.stage.Stage;
```

然后，在公共类 EmbeddedMediaPlayer 行之后添加 Example 6-2 的代码。

Example 6-2　Specify the Media File Source

```java
public class EmbeddedMediaPlayer extends Application {
private static final String MEDIA_URL = "http://flv3.people.com.cn/dev1/mvideo/vodfiles/2020/03/22/20c1218a_a16a9ccd505a04408b34990b_c.mp4";
```

（5）修改 start() 方法，使其如同 Example 6-3。这将创建一个具有组根节点和宽 540、高 210 的空场景。

Example 6-3　Modify the start Method

```java
@Override
public void start(Stage primaryStage) {
    primaryStage.setTitle("Embedded Media Player");
    Group root = new Group();
    Scene scene = new Scene(root, 540, 210);
    primaryStage.setScene(scene);
    primaryStage.sizeToScene();
    primaryStage.show();
}
```

（6）通过在 primaryStage 之前添加 Example 6-4 中的代码定义媒体和 MediaPlayer 对象，以设置场景的行。将 autoPlay 变量设置为 true，以使视频可以立即启动。

Example 6-4 Add media andmediaPlayer Objects

```
//创建 mediaPlayer 对象
Media media = new Media(MEDIA_URL);
MediaPlayer mediaPlayer = new MediaPlayer(media);
mediaPlayer.setAutoPlay(true);
```

（7）定义 MediaView 对象并将媒体播放器添加到基于节点的查看器中，方法是复制 Example 6-5 中的注释和两行代码，并将其粘贴到 mediaPlayer 之后。设置自动播放（真）行。

Example 6-5 Define MediaView Object

```
//创建 mediaView 对象并将 mediaPlayer 对象添加其中
MediaView mediaView = new MediaView(mediaPlayer);
((Group)scene.getRoot()).getChildren().add(mediaView);
```

（8）右击任何空白处，然后选择"格式"选项，以修复后的行格式添加代码行。

（9）在项目窗口中右击 EmbeddedMediaPlayer 项目节点，然后选择"清洁和构建"选项。

（10）成功构建后，右击项目节点运行应用程序，然后选择 Run 选项。

6.6　控制媒体播放

本节将创建一个功能齐全的媒体播放器，其中包含控制播放的图形用户界面元素。要创建媒体播放器，需要实现 3 个嵌套媒体对象的结构对图形控件进行编码，并为播放功能添加一些控制逻辑，如图 6.4 所示。

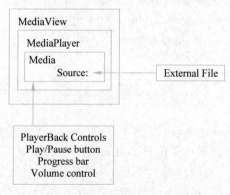

图 6.4　添加控制逻辑的媒体播放器的结构

添加的媒体控制面板由 3 个元素组成：播放按钮、进度和音量控制。

6.7 创建控件

本节将创建一个新的 JavaFX 文件 MediaControl.java，它将包含用于播放/暂停、进度和音量功能的窗口和 UI 控件。

（1）在 NetBeans IDE 中将 EmbeddedMediaPlayer 作为主项目打开，创建一个新的 JavaFX 文件并添加到项目中。

① 使用快捷键 Ctrl+N 或从 IDE 的主菜单中选择"文件"→"新建文件"选项。

② 选择类别 JavaFX 和文件类型 JavaFX 主类。

③ 在"名称和位置"对话框中，在"类名"字段中输入 MediaControl。

④ 在 Package 字段中，从下拉列表中选择 embeddedmediaplayer 选项，然后单击 Finish 按钮。

（2）在 MediaControl 中的 Java 源文件中，删除包 embeddedmediaplayer 行后的所有行。

（3）将 Example 6-6 中的导入语句添加到文件顶部。

Example 6-6　Import Statements to Add

```java
import javafx.scene.control.Label;
import javafx.scene.control.Slider;
import javafx.scene.layout.BorderPane;
import javafx.scene.layout.HBox;
import javafx.scene.layout.Pane;
import javafx.scene.media.MediaPlayer;
import javafx.scene.media.MediaView;
import javafx.util.Duration;
```

（4）复制并粘贴 Example 6-7 中的代码行，以创建控制按钮。

Example 6-7　Add MediaControl Class Code

```java
public class MediaControl extends BorderPane {
private MediaPlayer mp;
private MediaView mediaView;
private final boolean repeat = false;
private boolean stopRequested = false;
```

```
private boolean atEndOfMedia = false;
private Duration duration;
private Slider timeSlider;
private Label playTime;
private Slider volumeSlider;
private HBox mediaBar;
public MediaControl(final MediaPlayer mp) {
this.mp = mp;
setStyle("-fx-background-color: #bfc2c7;");
mediaView = new MediaView(mp);
Pane mvPane = new Pane() {
};
mvPane.getChildren().add(mediaView);
mvPane.setStyle("-fx-background-color: black;");
setCenter(mvPane);
}
}
```

（5）复制 Example 6-8 中的代码行,并将其粘贴到显示 setCenter(mvPane)的行之后。此代码添加了媒体工具栏和播放按钮。

Example 6-8 Add Media Toolbar and Play Button

```
mediaBar = new HBox();
mediaBar.setAlignment(Pos.CENTER);
mediaBar.setPadding(new Insets(5, 10, 5, 10));
BorderPane.setAlignment(mediaBar, Pos.CENTER);
final Button playButton = new Button(">");
mediaBar.getChildren().add(playButton);
setBottom(mediaBar);
}
}
```

（6）将 Example 6-9 中的导入语句添加到导入语句列表的顶部。

Example 6-9 Add More Import Statements

```
import javafx.geometry.Insets;
import javafx.geometry.Pos;
import javafx.scene.control.Button;
```

(7) 将剩余的 UI 控件添加到控制窗口中。将 Example 6-10 中的代码行放在 mediaBar 之后。

Example 6-10   Add the Rest of the UI Controls

```java
//添加 spacer
Label spacer = new Label(" ");
mediaBar.getChildren().add(spacer);
//添加 timeLabel 标签
Label timeLabel = new Label("Time: ");
mediaBar.getChildren().add(timeLabel);
//添加 timeslider 滚动条
timeSlider = new Slider();
HBox.setHgrow(timeSlider,Priority.ALWAYS);
timeSlider.setMinWidth(50);
timeSlider.setMaxWidth(Double.MAX_VALUE);
mediaBar.getChildren().add(timeSlider);
//添加 Play 标签
playTime = new Label();
playTime.setPrefWidth(130);
playTime.setMinWidth(50);
mediaBar.getChildren().add(playTime);
//添加 volume 标签
Label volumeLabel = new Label("Vol: ");
mediaBar.getChildren().add(volumeLabel);
//添加 Volume 滚动条
volumeSlider = new Slider();
volumeSlider.setPrefWidth(70);
volumeSlider.setMaxWidth(Region.USE_PREF_SIZE);
volumeSlider.setMinWidth(30);
mediaBar.getChildren().add(volumeSlider);
```

(8) 在文件顶部添加更多导入语句，如 Example 6-11 所示。

Example 6-11   Add More Import Statements

```java
import javafx.scene.layout.Priority;
import javafx.scene.layout.Region;
```

6.8 添加逻辑功能代码

创建所有控件并将其添加到控制面板后,添加功能逻辑以管理媒体播放,并使应用程序具有交互性。

(1) 为"播放"按钮添加事件处理程序和侦听器。将 Example 6-12 中的代码行复制并粘贴到最后一行按钮 playButton=new Button(">")之后以添加(播放按钮)行。

Example 6-12　Add Play Button's Event Handler and Listener

```
playButton.setOnAction(new EventHandler<ActionEvent>() {
    public void handle(ActionEvent e) {
        Status status = mp.getStatus();
        if (status == Status.UNKNOWN || status == Status.HALTED) {
            //在这些状态下不做任何事情
            return;
        }
        if ( status == Status.PAUSED || status == Status.READY || status == Status.STOPPED) {
            //如果播放结束
            if (atEndOfMedia) {
                mp.seek(mp.getStartTime());
                atEndOfMedia = false;
            }
            mp.play();
        } else
            mp.pause();
    }
});
```

(2) Example 6-12 中添加的代码使用的导入语句可以事先添加,以避免出错。但这一次,要消除所有标记的错误,按 Ctrl+Shift+I 键或右击任意位置并选择"修复导入"选项。在"修复所有导入"对话框中选择 javafx.scene.media.MediaPlayer.Status、javafx.event.ActionEvent 以及 javafx.event.EventHandler,然后单击 OK 按钮。

(3) 在 Example 6-10 中添加的代码行之后,在显示 mediaBar 的代码行之前添加以下代码行,该代码将处理侦听器。

Example 6-13　Add Listener Code

```
mp.currentTimeProperty( ).addListener(new InvalidationListener( ) {
    public void invalidated(Observable ov) {
        updateValues( );
    }
});
mp.setOnPlaying(new Runnable( ) {
    public void run( ) {
      if (stopRequested) {
        mp.pause( );
        stopRequested = false;
        } else {
        playButton.setText("||");
      }
    }
});
mp.setOnPaused(new Runnable( ) {
    public void run() {
    System.out.println("onPaused");
    playButton.setText(">");
}
});
mp.setOnReady(new Runnable( ) {
  public void run( ) {
  duration = mp.getMedia( ).getDuration( );
  updateValues( );
  }
});
mp.setCycleCount(repeat ? MediaPlayer.INDEFINITE : 1);
mp.setOnEndOfMedia(new Runnable( ) {
  public void run( ) {
  if (!repeat) {
    playButton.setText(">");
    stopRequested = true;
    atEndOfMedia = true;
  }
  }
});
```

注意：出现的错误将通过在接下来的步骤中添加更多的代码修复。

（4）通过在显示 timeSlider 的行后添加 Example 6-14 所示的代码为时间滑块添加侦听器。设置 MaxWidth(Double.MAX_值)，并在表示 mediaBar 的行之前添加（时间滑块）。

Example 6-14　Add Listener for Time Slider

```
timeSlider.valueProperty().addListener(new InvalidationListener(
) {
    public void invalidated(Observable ov) {
      if (timeSlider.isValueChanging()) {
      //将持续时间乘以滑块位置计算的百分比
      mp.seek(duration.multiply(timeSlider.getValue() / 100.0));
      }
    }
});
```

（5）通过在 volumeSlider 一行之后添加 Example 6-15 所示的代码为音量滑块控件添加侦听器。

Example 6-15　Add Listener for the Volume Control

```
volumeSlider.valueProperty().addListener(new InvalidationListener
() {
    public void invalidated(Observable ov) {
      if (volumeSlider.isValueChanging()) {
       mp.setVolume(volumeSlider.getValue() / 100.0);
      }
    }
});
```

（6）创建播放控件使用的方法 UpdateValue( )，如 Example 6-16 所示。将其添加到 public MediaControl( )方法之后。

Example 6-16　Add UpdateValues Method

```
protected void updateValues() {
    if (playTime != null && timeSlider != null && volumeSlider != null) {
       Platform.runLater(new Runnable() {
```

```
        public void run( ) {
            Duration currentTime = mp.getCurrentTime( );
            playTime.setText(formatTime(currentTime, duration));
            timeSlider.setDisable(duration.isUnknown( ));
            if (!timeSlider.isDisabled()&&duration.greaterThan
(Duration.ZERO)&&!timeSlider.isValueChanging( )) {
                timeSlider.setValue(currentTime.divide(duration).
toMillis( ) * 100.0);
            }
            if (!volumeSlider.isValueChanging( )) {
                volumeSlider.setValue((int)Math.round(mp.getVolume( ) *
100));
            }
        }
    });
  }
}
```

(7) 在 updateValues()方法之后添加私有方法 formatTime()，该方法可以计算媒体播放所用的时间，并将其格式化以显示在控件工具栏上。

Example 6-17 Add Method for Calculating Elapsed Time

```
private static String formatTime ( Duration elapsed, Duration
duration) {
    int intElapsed = (int)Math.floor(elapsed.toSeconds( ));
    int elapsedHours = intElapsed / (60 * 60);
    if (elapsedHours > 0) {
      intElapsed -= elapsedHours * 60 * 60;
    }
    int elapsedMinutes = intElapsed / 60;
     int elapsedSeconds = intElapsed - elapsedHours * 60 * 60 -
elapsedMinutes * 60;
    if (duration.greaterThan(Duration.ZERO)) {
      int intDuration = (int)Math.floor(duration.toSeconds( ));
      int durationHours = intDuration / (60 * 60);
      if (durationHours > 0) {
        intDuration -= durationHours * 60 * 60;
      }
```

```
            int durationMinutes = intDuration / 60;
            int durationSeconds = intDuration - durationHours * 60 * 60 -
durationMinutes * 60;
            if (durationHours > 0) {
                return String. format ( "% d:% 02d:% 02d/% d:% 02d:% 02d",
elapsedHours, elapsedMinutes, elapsedSeconds,
durationHours, durationMinutes, durationSeconds);
            } else {
              return String.format("%02d:%02d/%02d:%02d", elapsedMinutes,
elapsedSeconds,durationMinutes,
durationSeconds);
            }
        } else {
            if (elapsedHours > 0) {
                return String. format ( "% d:% 02d:% 02d", elapsedHours,
elapsedMinutes, elapsedSeconds);
            } else {
                return String. format ( "% 02d:% 02d", elapsedMinutes,
elapsedSeconds);
            }
        }
    }
```

6.9 修改 EmbeddedMediaPlayer.java

要添加控件,需要请修改 EmbeddedMediaPlayer。

(1) 复制 Example 6-18 中的代码行,并将其粘贴到 mediaPlayer 后面,设置自动播放(真)行。

Example 6-18　Add the Source Code to Create MediaControl Object

```
MediaControl mediaControl = new MediaControl(mediaPlayer);
scene.setRoot(mediaControl);
```

(2) 删除 Example 6-19 中显示的 3 行,这 3 行之前创建了 mediaView 和 mediaPlayer 对象。

第 6 章 JavaFX Media 应用开发

Example 6-19 Delete Lines of Code

```
//创建 mediaView 对象并将其添加到 mediaView 中
MediaView mediaView = new MediaView(mediaPlayer);
((Group)scene.getRoot()).getChildren().add(mediaView);
```

（3）删除 MediaView：import 的导入语句"import javafx.scene.media.MediaView;"。

（4）调整场景高度的大小，以适应媒体的添加控制。

Example 6-20 Change the Scene's Height

```
Scene scene = new Scene(root, 540, 241);
```

编译并运行 EmbeddedMedia，现在构建在 6.8 节中创建的应用程序并运行它。

（1）右击 EmbeddedMediaPlayer 项目节点，然后选择 Clean and Build 选项。

（2）如果没有生成错误，则再次右击该节点并选择"运行"选项，运行结果如图 6.5 所示。

图 6.5 媒体播放器运行结果

EmbeddedMediaPlayer.java 的完整源代码如下所示。

```
package embeddedmediaplayer;
import javafx.application.Application;
import javafx.scene.Group;
import javafx.scene.Scene;
import javafx.scene.media.Media;
import javafx.scene.media.MediaPlayer;
```

```java
import javafx.scene.media.MediaView;
import javafx.stage.Stage;
public class EmbeddedMediaPlayer extends Application {
    private static final String MEDIA_URL = "http://flv3.people.com.cn/dev1/mvideo/vodfiles/2020/03/22/20c1218aa16a9ccd505a04408b34990b_c.mp4";
    @Override
    public void start(Stage primaryStage) {
        primaryStage.setTitle("Embedded Media Player");
        Group root = new Group();
        Scene scene = new Scene(root, 540, 210);
        //创建 media 对象
        Media media = new Media(MEDIA_URL);
        MediaPlayer mediaPlayer = new MediaPlayer(media);
        mediaPlayer.setAutoPlay(true);
        //创建 MediaView 对象并将其添加到 Viewer 中
        MediaView mediaView = new MediaView(mediaPlayer);
        ((Group) scene.getRoot()).getChildren().add(mediaView);
        primaryStage.setScene(scene);
        primaryStage.sizeToScene();
        primaryStage.show();
    }
    public static void main(String[] args) {
        launch(args);
    }
}
```

6.10 小结

随着 Internet 上媒体内容的持续增长，视频和音频已经成为富 Internet 应用的重要组成部分。如果充分利用 JavaFX 的多媒体功能，就能够极大地拓宽传统媒体使用范围。本章通过一个综合示例分步骤详细介绍了 JavaFX Media 程序设计方面的知识。

第 7 章 JavaFX 3D 应用开发

本章将介绍 JavaFX 图形开发技术，包括 JavaFX 3D 图形入门、使用图形操作 API 以及使用 Canvas API。每部分都提供了 JavaFX 图形功能的可用 API 信息，包括代码示例和如何使用 API 进行 3D 应用开发。JavaFX 的 3D 图形部分包含以下知识点：Shape 3D、Camera、SubScene、Light、Material、Picking 等。具备这些基础知识后，本章将引导读者构建一个综合的示例应用。

7.1 Shape 3D

本节介绍有关 JavaFX 3D 图形库中提供的 Shape 3D API 的信息。Shape 3D API 提供了两种类型的 3D 形状：预定义形状与用户定义形状。

1. 预定义形状

运用预定义 3D 形状，开发人员能够快速创建 3D 对象，这些形状包括长方体、圆柱体和球体，Shape 3D 类的层次结构如 Example 7-1 所示，它包含 MeshView 类，该类使用指定的三维网格数据实现了曲面的定义，还包括长方体、圆柱体和球体子类。

Example 7-1　Shape 3D Class Hierarchy

```
java.lang.Object
javafx.scene.Node
javafx.scene.shape.Shape3D
javafx.scene.shape.MeshView
javafx.scene.shape.Box
javafx.scene.shape.Cylinder
javafx.scene.shape.Sphere
```

应用以下信息可以创建预定义形状。
（1）创建长方体对象需要指定宽度、高度以及深度的尺寸。

```
Box myBox = new Box(width, height, depth);
```

(2) 创建圆柱体对象需要指定半径和高度。

```
Cylinder myCylinder = new Cylinder(radius, height);
Cylinder myCylinder2 = new Cylinder(radius, height, divisions);
```

(3) 创建球体对象需要指定半径。

```
Sphere mySphere = new Sphere(radius);
Sphere mySphere2 = new Sphere(radius, divisions);
```

Example 7-2 为预定义 3D 形状使用的代码行。

Example 7-2　Sample Usage of Predefined 3D Shapes

```
...
final PhongMaterial redMaterial = new PhongMaterial();
redMaterial.setSpecularColor(Color.ORANGE);
redMaterial.setDiffuseColor(Color.RED);
final PhongMaterial blueMaterial = new PhongMaterial();
blueMaterial.setDiffuseColor(Color.BLUE);
blueMaterial.setSpecularColor(Color.LIGHTBLUE);
final PhongMaterial greyMaterial = new PhongMaterial();
greyMaterial.setDiffuseColor(Color.DARKGREY);
greyMaterial.setSpecularColor(Color.GREY);
final Box red = new Box(400, 400, 400);
red.setMaterial(redMaterial);
final Sphere blue = new Sphere(200);
blue.setMaterial(blueMaterial);
final Cylinder grey = new Cylinder(5, 100);
grey.setMaterial(greyMaterial);
...
```

2. 用户定义形状

Example 7-3 显示了 JavaFX 网格类的层次结构,其中包含 TriangleMesh 子类。三角形网格是 3D 布局中最典型的网格类型。

Example 7-3　Mesh Class Hierarchy

```
java.lang.Object
```

```
javafx.scene.shape.Mesh (abstract)
javafx.scene.shape.TriangleMesh
```

三角形网格包含描述三角化几何网格的点、纹理坐标和面的单独阵列。平滑组用于对属于同一曲面的三角形进行分组。不同平滑组中的三角形可以形成硬边。

可以按以下步骤创建三角形网格实例。

（1）创建三角形网格的实例对象。

```
mesh = new TriangleMesh();
```

（2）定义点集，即网格的顶点。

```
float points[ ] = { … };
mesh.getPoints().addAll(points);
```

（3）描述每个顶点的纹理坐标。

```
float texCoords[ ] = { … };
mesh.getTexCoords().addAll(texCoords);
```

（4）使用顶点构建面，这些面是描述拓扑的三角形。

```
int faces[ ] = { … };
mesh.getFaces().addAll(faces);
```

（5）定义每个面所属的平滑组。

```
int smoothingGroups[ ] = { … };
mesh.getFaceSmoothingGroups().addAll(smoothingGroups);
```

平滑组调整顶点上的法线，使面光滑或多面。如果每个面都有不同的平滑组，那么网格将是多面的；如果每个面都有相同的平滑组，则网格将非常平滑。

7.2 Camera 3D

本节将介绍 JavaFX 3D 图形附带的相机 API 的特征。相机是一个节点，可以添加到 JavaFX 程序的场景图中，从而允许开发人员在 3D UI 布局中移动相机。而在 2D UI 布局中,相机只能保持在一个位置上的布局。在 JavaFX 场景坐标空间中,默认相机的投影平面为 Z=0 的相机坐标系的定义如下：
- X 轴指向右侧；
- Y 轴指向下方；
- Z 轴指向远离观众或进入屏幕的方向。

1. 透视相机

JavaFX 提供了用于渲染 3D 场景的透视相机,该相机定义了透视投影的观察体积。可以通过更改"视野"属性的值更改查看体积。Example 7-4 显示了用于创建透视相机的两个构造方法。

Example 7-4　Constructors for PerspectiveCamera

```
PerspectiveCamera()
PerspectiveCamera(boolean fixedEyeAtCameraZero)
```

第二个构造方法是 JavaFX 8 中的一个新的构造方法,它允许使用指定的 FixedYeatCameraZero 标志控制相机的位置,以便渲染相机在 3D 环境中看到的内容。

3D 图形编程一般使用以下的造方法：

```
PerspectiveCamera(true);
```

当选项 FixedYeatCamerazero 属性设置为 true 时,透视相机的眼睛位置在其坐标空间中固定在(0,0,0),而不管投影区域的尺寸或窗口大小的变化。当 FixedYeatCamerazero 属性设置为默认值 false 时,相机定义的坐标系的原点位于面板的左上角,该模式用于使用透视相机渲染的 2D UI 控件,但对大多数 3D 图形应用来说很少使用。调整窗口大小时,相机也会随之移动,例如将原点保持在面板的左上角。这正是 2D UI 布局所需要的,但 3D 布局并不需要。因此,在进行 3D 图形变换或者移动相机时,一定要记住将 FixedYeatCamerazero 属性设置为 true。

创建相机并将其添加到场景中,可以使用以下代码：

```
Camera camera = new PerspectiveCamera(true);
scene.setCamera(camera);
```

用以下代码将相机添加到场景图中。

```
Group cameraGroup = new Group();
```

```
cameraGroup.getChildren().add(camera);
root.getChildren().add(cameraGroup);
```

旋转相机并移动 cameraGroup，可以用以下代码：

```
camera.rotate(45);
cameraGroup.setTranslateZ(-75);
```

2. 视野

相机的视野可以设置如下：

```
camera.setFieldOfView(double value);
```

视野越大，透视失真和尺寸的差异越大。
- 鱼眼镜头的视野可达 180°及以上。
- 普通透镜的视野在 40°~62°之间。
- 长焦镜头的视野为 1°(或更小)~30°。

3. 裁剪平面

可以在局部坐标系中设置相机的近剪裁平面，如下所示：

```
camera.setNearClip(double value);
```

可以在局部坐标系中设置相机的远剪裁平面，如下所示：

```
camera.setFarClip(double value);
```

设置近剪裁平面或远剪裁平面可以确定查看体积。如果近剪裁平面太大，则它基本上会开始剪裁场景的前部。如果太小，则会开始剪裁场景的后部。

注意：不要将"近剪裁"值设置为小于所需的值，或将"远剪裁"值设置为大于所需的值，否则可能会出现奇怪的视觉瑕疵。

需要设置剪裁平面，以便可以看到足够多的场景。但是查看范围不应设置得太大，以免出现数字错误。如果近剪裁平面的值太大，则场景开始被剪裁；如果附近剪切平面太小，则会因某个值而出现另一种视觉伪影太接近于 0；如果远剪裁平面的值太大，也会出现数值错误，尤其是在近剪裁平面的值太小的情形下。

4. Y-down 与 Y-up

大多数二维图形坐标系(包括用户界面)的 Y 轴会随着时间的增加出现在屏幕下面。Photoshop、JavaFX 和 Illustrator 也是如此,基本上大部分 2D 软件包都是这样工作的。许多三维图形坐标系通常具有 Y 轴随着屏幕的上移而增加的功能。一些三维图形坐标系的 Z 轴会随着用户向上移动而增加,而且大多数的情形是随着用户向上移动屏幕而增加。

Y 向下和 Y 向上在各自的语境中都是正确的。在 JavaFX 中,相机的坐标系是 Y 向下的,这意味着 X 轴指向右侧,Y 轴指向右侧向下,Z 轴指向远离用户或屏幕的方向。

如果希望将 3D 场景视为 Y 向上的,则可以创建一个 root3D,如 Example 7-5 所示,并设定它的 rx。将属性设置为 180°,基本上把它颠倒过来。然后,将 3D 元素添加到 root3D 节点中,并将相机置于 root3D 下。

Example 7-5 Create Xform node,root3D

```
root3D = new Xform();
root3D.rx.setAngle(180.0);
root.getChildren().add(root3D);
root3D.getChildren().add(...); //add all your 3D nodes here
```

还可以创建一个名为 cameraXform 的 Xform 节点,并将其放在根目录下,如 Example 7-6 所示。把它颠倒过来,把相机放在 cameraXform 下面。

Example 7-6 Create a cameraXform Node

```
Camera camera = new PerspectiveCamera(true);
Xform cameraXform = new Xform();
root.getChildren().add(cameraXform);
cameraXform.getChildren().add(camera);
cameraXform.rz.setAngle(180.0);
```

一个更好的方法是在相机节点上添加一个 180°的旋转,这是一个细微的差异。使用的旋转不是为用户提供的旋转,因为用户希望避免自动旋转。在 Example 7-7 中,将相机旋转 180°,然后将其作为 cameraXform 的子对象添加到相机中。区别在于,cameraXform 保留了原始的值,在其默认位置,所有内容都将归零,包括平移和旋转。

Example 7-7 Create cameraXform and Rotate

```
Camera camera = new PerspectiveCamera(true);
```

```
Xform cameraXform = new Xform();
root.getChildren().add(cameraXform);
cameraXform.getChildren().add(camera);
Rotate rz = new Rotate(180.0, Rotate.Z_AXIS);
camera.getTransforms().add(rz);
```

5. 使用 PerspectiveCamera 的示例代码

Example 7-8 中显示的 Simple3DBoxApp 创建了一个 3D 框,并使用透视相机渲染场景。此示例应用程序如下所示。

Example 7-8　3D Box Sample Application

```
package simple3dbox;
import javafx.application.Application;
import javafx.scene.Group;
import javafx.scene.Parent;
import javafx.scene.PerspectiveCamera;
import javafx.scene.Scene;
import javafx.scene.SubScene;
import javafx.scene.paint.Color;
import javafx.scene.paint.PhongMaterial;
import javafx.scene.shape.Box;
import javafx.scene.shape.DrawMode;
import javafx.scene.transform.Rotate;
import javafx.scene.transform.Translate;
import javafx.stage.Stage;
public class Simple3DBoxApp extends Application {
  public Parent createContent() throws Exception {
    Box testBox = new Box(5, 5, 5);
    testBox.setMaterial(new PhongMaterial(Color.RED));
    testBox.setDrawMode(DrawMode.LINE);
    //创建并定位相机
    PerspectiveCamera camera = new PerspectiveCamera(true);
    camera.getTransforms().addAll(new Rotate(-20, Rotate.Y_AXIS),
new Rotate(-20, Rotate.X_AXIS), new Translate(0, 0, -15));
    //构建场景图
    Group root = new Group();
    root.getChildren().add(camera);
    root.getChildren().add(testBox);
```

```java
    //使用SubScene
    SubScene subScene = new SubScene(root, 300,300);
    subScene.setFill(Color.ALICEBLUE);
    subScene.setCamera(camera);
    Group group = new Group();
    group.getChildren().add(subScene);
    return group;
}
@Override
public void start(Stage primaryStage) throws Exception {
    primaryStage.setResizable(false);
    Scene scene = new Scene(createContent());
    primaryStage.setScene(scene);
    primaryStage.show();
}
////通过调用launch()方法启动JavaFX应用
public static void main(String[] args) {
    launch(args);
}
}
```

【运行结果】

执行结果如图 7.1 所示。

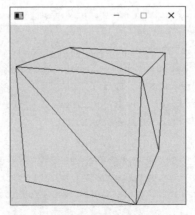

图 7.1 Simple3DBoxApp 的运行结果

7.3 SubScene

本节介绍有关在 JavaFX 中使用 SubScene API 的信息。子场景节点是场景图中内容的容器，这是一个特殊的节点场景分离，它可以使用不同的相机渲染场景的一部分。如果希望三维对象具有 Y 向上，则可以使用子场景节点布局中的二维 UI 对象。一些可能的子场景用例包括：

- UI 控件的覆盖（需要一个静态相头）；
- 背景底图（静态或更新频率较低）；
- "抬头"显示；
- Y 向上表示 3D 对象，Y 向下表示 2D UI。

1. 创建一个 SubScene

Example 7-9 显示了在应用程序中创建子场景节点新实例对象的两个构造方法。

Example 7-9　SubScene Constructors

```
//为具有特定大小的特定根节点创建子场景
public SubScene(Parent root, double width, double height)
//构造一个由根组成的子场景，其维度为宽度和高度，指定是否为此场景创建深度缓
//冲区，并指定是否请求场景抗锯齿
public SubScene (Parent root, double width, double height, boolean
depthBuffer, SceneAntialiasing antiAliasing)
```

创建子场景后，可以使用方法对其进行修改，以指定或获取子场景的高度、根节点、宽度、背景填充、渲染子场景的相机类型，或者子场景是否抗锯齿。

2. 子场景的示例

CreateSubScene( )方法（如 Example 7-10 所示）展示了如何使用上面列出的第二个子场景的构造方法。

Example 7-10　Code Sample Using SubScene

```
...
SubScene msaa = createSubScene ("MSAA = true", cylinder2, Color.
TRANSPARENT, new PerspectiveCamera(), true);
...
private static SubScene createSubScene(String title, Node node, Paint
fillPaint, Camera camera, boolean msaa) {
```

```
    Group root = new Group();
    PointLight light = new PointLight(Color.WHITE);
    light.setTranslateX(50);
    light.setTranslateY(-300);
    light.setTranslateZ(-400);
    PointLight light2 = new PointLight(Color.color(0.6, 0.3, 0.4));
    light2.setTranslateX(400);
    light2.setTranslateY(0);
    light2.setTranslateZ(-400);
    AmbientLight ambientLight = new AmbientLight(Color.color(0.2, 0.2,
0.2));
    node.setRotationAxis(new Point3D(2, 1, 0).normalize());
    node.setTranslateX(180);
    node.setTranslateY(180);
    root.getChildren().addAll(setTitle(title), ambientLight, light,
light2, node);
     SubScene  subScene = new  SubScene ( root, 500, 400, true, msaa?
SceneAntialiasing.BALANCED :SceneAntialiasing.DISABLED);
    subScene.setFill(fillPaint);
    subScene.setCamera(camera);
    return subScene;
}
```

在图形 3D 部分中，提供的 3D 立方体和木琴样本也说明了 SubScene API 的使用。

7.4　Light

本节介绍 JavaFX 3D 图形库中包含的 Light API。灯光被定义为场景图中的节点，如果场景中包含的活动灯光集为空，则会提供默认灯光。每个灯光都包含一组受影响的节点。如果一组节点为空，则场景（或子场景）上的所有节点都会受到影响。如果父节点位于该节点中，则其所有子节点也会受到影响。灯光与 Shape 3D 对象的几何体及其材质交互，以提供渲染效果。目前，有以下两种类型的光源：

- 环境光——来自四面八方的光源；
- 点光源——在空间中有一个固定点并辐射光的光源，在远离自身的所有方向上都是平等的。

Example 7-11 显示了 Light 类的层次结构。

Example 7-11　Light Class Hierarchy

```
javafx.scene.Node
javafx.scene.LightBase (abstract)
javafx.scene.AmbientLight
javafx.scene.PointLight
```

要创建点光源并将其添加到场景中,需要执行如下操作:

```
PointLight light = new PointLight();
light.setColor(Color.RED);
```

使用如下命令将灯光添加到场景图中:

```
Group lightGroup = new Group();
lightGroup.getChildren().add(light);
root.getChildren().add(lightGroup);
```

输入如下代码将灯旋转 45°:

```
light.rotate(45);
```

要移动灯光组并使灯光随之移动,可以使用如下代码:

```
lightGroup.setTranslateZ(-75);
```

setTranslateZ( )方法用于设置 translateZ 属性的值,在上面的示例代码中,该属性的值为 −75。该值将添加到 transforms ObservableList 和 layoutZ 定义的任何转换中。

Example 7-12 显示了取自 MSAAPP.java 应用程序的代码片段,它演示了如何使用 PointLight API。

Example 7-12　MSAAApp.java Code Snippet Using PointLight API

```
PointLight light = new PointLight(Color.WHITE);
light.setTranslateX(50);
light.setTranslateY(-300);
light.setTranslateZ(-400);
PointLight light2 = new PointLight(Color.color(0.6, 0.3, 0.4));
```

```
light2.setTranslateX(400);
light2.setTranslateY(0);
light2.setTranslateZ(-400);
AmbientLight ambientLight = new AmbientLight(Color.color(0.2, 0.2,
0.2));
node.setRotationAxis(new Point3D(2, 1, 0).normalize());
node.setTranslateX(180);
node.setTranslateY(180);
root.getChildren().addAll(setTitle(title), ambientLight, light,
light2, node);
```

7.5 Material

本节将介绍 JavaFX 3D 图形库的材质类——Material。Material 类包含一组渲染属性。Example 7-13 显示了材质类的层次结构，并且 PhongMaterial 类是从材质类中划分出来的子类。

Example 7-13 Material Class Hierarchy

```
java.lang.Object
javafx.scene.paint.Material (abstract)
javafx.scene.paint.PhongMaterial
```

PhongMaterial 类提供了表示 Phong 着色材质形式的特性定义：
- Diffuse color——漫反射颜色；
- Diffuse map——漫反射贴图；
- Specular map——镜面反射贴图；
- Specular color——镜面反射颜色；
- Specular power——镜面反射功率；
- Bump map or normal map——凹凸贴图或法线贴图；
- Self-illumination map——自发光贴图。

Material 类可在多个 Shape 3D 节点之间共享。Example 7-14 显示了如何创建 PhongMaterial 对象，以及设置其扩散贴图属性，并将材质用于形状。

Example 7-14 Working with Material

```
Material mat = new PhongMaterial();              //创建 Material 对象
Image diffuseMap = new Image("diffuseMap.png");
```

```
Image normalMap = new Image("normalMap.png");
//设置 material 属性
mat.setDiffuseMap(diffuseMap);
mat.setBumpMap(normalMap);
mat.setSpecularColor(Color.WHITE);
shape3d.setMaterial(mat);                    //将 Material 用于形状
```

7.6 Picking

本节介绍 JavaFX 3D 附带的 PickResult API 的图形功能。PickResult API 可用于带有透视图的 2D 相机。然而，由于与深度缓冲区一起使用时存在局限性，因此 PickResult 类已添加到 javafx.scene.input 包中，它是一个包含 pick 事件结果的容器对象。

PickResult 参数已添加到 MouseEvent 的所有构造方法中，MousedRageEvent、DrageEvent、GestureEvent、ContextMenuEventy 以及 TouchPoint 类均可以返回有关用户选择的信息。新增的这些类中的 getPickResult( )方法用来返回一个新的 PickResult 对象，该对象包含关于选择的信息。添加的另一个方法是 getZ( )，它返回深度事件相对于 MouseEvent 源的起源位置。

1. 创建 PickResult 对象

JavaFX API 提供了 3 个构造方法，用于在应用中创建 PickResult 对象实例，如 Example 7-15 所示。

为了不借助其他信息的二维案例创建 Pick 结果，将给定的场景坐标转换为目标的局部坐标空间，并将该值存储为交点。将相交节点设置为给定目标，距离设置为 1.0，texCoord 设置为空。

Example 7-15　PickResult Constructors

```
PickResult(EventTarget target, double sceneX, double sceneY)
```

为非 3D 形状目标创建 PickResult 的新实例对象。将 face to face_UNDEFINED 和 texCoord 设置为 null。

```
PickResult(Node node, Point3D point, double distance)
```

创建一个新的 PickResult 实例对象。

```
PickResult(Node node, Point3D point, double distance, int face,
Point2D texCoord)
```

2. PickResult 对象的方法

在代码中创建 PickResult 对象后,可以使用以下方法处理从处理事件的类所传递的信息。

Example 7-16   PickResult Methods

```
//返回相交的节点。如果没有与任何节点相交且已拾取场景,则返回 null
public final Node getIntersectedNode();
//返回拾取节点的局部坐标中的交点。如果未拾取任何节点,则返回与子平面相交
//的点
public final Point3D getIntersectedPoint();
//返回相机位置和相交点之间的相交距离
public final double getIntersectedDistance();
//返回拾取节点的相交面,如果节点没有用户指定的面或是在边界上拾取的,则返回
//未定义的面
public final int getIntersectedFace();
//返回拾取的三维形状的相交纹理坐标。如果拾取的目标不是 Shape3D 或
//pickOnBounds==true,则返回 null
//返回表示相交 TexCoord 的新 Point2D
public final Point2D getIntersectedTexCoord();
```

3. PickResult 的示例使用

Example 7-17 展示了如何使用 PickResult 对象及其方法。这些代码片段是 PickMesh3DSample 应用程序的一部分,该应用程序演示了如何访问 PickResult 对象中的信息。将光标悬停在网格上时,有关光标位置的信息将显示在 Overlay 中。可以按 L 键在"填充"和"线"之间切换绘制模式,以查看形成网格的每个面。

Example 7-17   Code Sample Using PickResult

```
...
EventHandler<MouseEvent> moveHandler = new EventHandler<MouseEvent
>() {
    @Override
    public void handle(MouseEvent event) {
      PickResult res = event.getPickResult();
```

```
      setState(res);
      event.consume();
    }
    ...
    final void setState(PickResult result) {
      if (result.getIntersectedNode() == null) {
          data.setText ( " Scene \ n \ n " + point3DToString (result.
getIntersectedPoint ( )) + " \ n " + point2DToString ( result.
getIntersectedTexCoord()) + "\n" + result.getIntersectedFace() + "\
n" + String.format("%.1f", result.getIntersectedDistance()));
      }
      else {
         data.setText(result.getIntersectedNode().getId() + "\n" +
getCullFace(result.getIntersectedNode()) + "\n" + point3DToString
(result.getIntersectedPoint ( )) + "\n" + point2DToString (result.
getIntersectedTexCoord()) + "\n" + result.getIntersectedFace() + "\
n" + String.format("%.1f", result.getIntersectedDistance()));
      }
    }
    ...
```

7.7 构建 3D 示例应用程序

本节阐述构建 MoleculeSampleApp 应用程序的步骤，该应用程序使用了前面讨论的一些 JavaFX 3D 图形的功能。本节中的操作步骤将基于 NetBeans IDE 13.0 进行开发。

本示例包括以下文件和附录 1 部分。

- MoleculeSampleApp.zip——MoleculeSampleApp 应用程序已完成的 NetBeans 项目。
- Xform.java——声明 Xform 类的方法。
- buildMolecule( )——创建三维水分子对象。
- handleMouse( ) and handleKeyboard( )——允许使用鼠标和键盘在场景中操纵相机的视图。

可以按照以下步骤进行开发：构建开发运行环境、创建项目、创造场景、调整相机、建造轴线、构建分子、添加照相头查看控件等。

1. 构建开发运行环境

在继续该操作之前,需要注意以下要求和建议。

(1) 确保使用的系统符合认证系统配置页面的要求,该链接来自 Java SE 下载页面 http://www.oracle.com/technetwork/java/javase/downloads/。

(2) JavaFX 图形支持部分提供了支持 JavaFX 3D 的特性。计算机中需要有图形卡的支持,才能成功运行构建的完整 JavaFX 3D 示例应用程序。

(3) 下载并安装 NetBeans IDE 1.3.0,以构建 JavaFX 3D 示例应用程序。

2. 创建 NetBeans 项目

通过以下步骤基于 NetBeans IDE 13.0 创建 MolecleSampleApp 的 JavaFX 项目。

(1) 在 NetBeans IDE 的"文件"菜单中选择"新建项目"选项。

(2) 在新建项目向导中选择 JavaFX 应用程序类别和 JavaFX 应用程序的项目,然后单击"下一步"按钮。

(3) 输入 MoleculeSampleApp 作为项目名称。输入路径的"位置"或单击"浏览"按钮导航到这个项目。最后单击 Finish 按钮。创建 JavaFX 项目时,NetBeans IDE 提供了一个 Hello World 源代码代码模板作为起点,将在下一节替换该模板中的源代码。

3. 创造场景

(1) 复制 Xform.java 文件内容,并将其保存到文件 Xform.java 中,该文件将保存在 moleculesampleapp 项目的 moleculesampleapp 源文件夹中。该文件包含派生自 Xform 子类的源代码。使用"变换"节点可以防止自动重新计算。更改组节点的子节点时,组节点轴心的位置在 3D UI 布局中。Xform 节点允许添加自己类型的变换和旋转。该文件包含一个平移组件,3 个旋转组件和比例组件。3 个旋转组件非常重要,它们在频繁更改旋转值时很有用,例如在更改旋转角度时。

(2) 在 IDE 编辑器中打开 MoleculeSampleApp.java 文件,替换页面顶部的导入语句,包含导入语句的文件如 Example 7-18 所示。

Example 7-18   Replacement Import Statements

```
import javafx.application.Application;
import javafx.scene.*;
import javafx.scene.paint.Color;
import javafx.stage.Stage;
```

(3) 替换 MolecleSampleApp 中代码主体其余部分的 Java 代码，如 Example 7-19 所示。该代码创建了一个新的场景图，其中一个 Xform 作为其节点。

Example 7-19　Replacement Body of Code

```java
public class MoleculeSampleApp extends Application {
    final Group root = new Group();
    final Xform world = new Xform();
    @Override
    public void start(Stage primaryStage) {
        Scene scene = new Scene(root, 1024, 768, true);
        scene.setFill(Color.GREY);
        primaryStage.setTitle("Molecule Sample Application");
        primaryStage.setScene(scene);
        primaryStage.show();
    }
    public static void main(String[] args) {
        launch(args);
    }
}
```

(4) 保存文件。

4. 设置相机

在带有 Xform 实例的类层次结构中设置相机。平移和旋转相机以更改其默认位置。

(1) 添加下面的代码，如 Example 7-20 所示。这些代码创建了一个 perspectiveCamera 的实例对象和 3 个实例公共类 Xform 的一部分，它扩展了 Group 类。在 7.6 节中添加到 NetBeans 项目中的 Java 文件是本文件的一部分。

Example 7-20　Add Variables for the Camera

```java
final Group root = new Group();
final Xform world = new Xform();
final PerspectiveCamera camera = new PerspectiveCamera(true);
final Xform cameraXform = new Xform();
final Xform cameraXform2 = new Xform();
final Xform cameraXform3 = new Xform();
private static final double CAMERA_INITIAL_DISTANCE = -450;
```

```
private static final double CAMERA_INITIAL_X_ANGLE = 70.0;
private static final double CAMERA_INITIAL_Y_ANGLE = 320.0;
private static final double CAMERA_NEAR_CLIP = 0.1;
private static final double CAMERA_FAR_CLIP = 10000.0;
```

(2) 复制 buildCamera( )方法的代码,如 Example 7-21 所示,将它们添加到变量声明的行之后。

buildCamera( )方法将相机设置为视图倒置,而不是默认的 JavaFX 2D 的 Y 倒置。因此,该场景被视为 Y 向上(Y 轴指向上)的场景。

Example 7-21　Add the buildCamera( ) Method

```
private void buildCamera() {
    root.getChildren().add(cameraXform);
    cameraXform.getChildren().add(cameraXform2);
    cameraXform2.getChildren().add(cameraXform3);
    cameraXform3.getChildren().add(camera);
    cameraXform3.setRotateZ(180.0);
    camera.setNearClip(CAMERA_NEAR_CLIP);
    camera.setFarClip(CAMERA_FAR_CLIP);
    camera.setTranslateZ(CAMERA_INITIAL_DISTANCE);
    cameraXform.ry.setAngle(CAMERA_INITIAL_Y_ANGLE);
    cameraXform.rx.setAngle(CAMERA_INITIAL_X_ANGLE);
}
```

(3) 在 start( )方法中添加对 buildCamera( )的调用,如 Example 7-22 所示。

Example 7-22　Add Method Call to buildCamera( )

```
public void start(Stage primaryStage) {
    buildCamera();
}
```

(4) 通过复制 Example 7-23 中的代码在场景中设置相机,并将其添加到 start( )方法的末尾。

Example 7-23　Set the Camera in the Scene

```
primaryStage.show();
scene.setCamera(camera);
```

(5）保存文件。

5．建造轴线

添加用于构建该分子的 3D 轴。Box 类用于创建轴，PhongMaterial 用于设置镜面反射和漫反射颜色。在 JavaFX 中，默认 Y 轴向下。按照通常的惯例，X 轴显示为红色，Y 轴显示为绿色，Z 轴显示为蓝色。

（1）将以下示例中的语句添加到导入的源文件中。

Example 7-24　Add Two Additional Import Statements

```java
import javafx.scene.paint.PhongMaterial;
import javafx.scene.shape.Box;
```

（2）添加下面的变量声明语句，如 Example 7-25 所示。

Example 7-25　Add a Variable for the Axes

```java
private static final double AXIS_LENGTH = 250.0;
```

（3）复制 Example 7-26 中的声明，并将其添加到声明 root 的行之后。

Example 7-26　Create the axisGroup

```java
final Group root = new Group();
final Xform axisGroup = new Xform();
```

（4）将 Example 7-27 中的 buildAxis( ) 方法添加到 buildCamera( ) 方法之后。

Example 7-27　Add buildAxes( ) Method

```java
private void buildAxes() {
    final PhongMaterial redMaterial = new PhongMaterial();
    redMaterial.setDiffuseColor(Color.DARKRED);
    redMaterial.setSpecularColor(Color.RED);
    final PhongMaterial greenMaterial = new PhongMaterial();
    greenMaterial.setDiffuseColor(Color.DARKGREEN);
    greenMaterial.setSpecularColor(Color.GREEN);
    final PhongMaterial blueMaterial = new PhongMaterial();
    blueMaterial.setDiffuseColor(Color.DARKBLUE);
    blueMaterial.setSpecularColor(Color.BLUE);
    final Box xAxis = new Box(AXIS_LENGTH, 1, 1);
    final Box yAxis = new Box(1, AXIS_LENGTH, 1);
```

```
        final Box zAxis = new Box(1, 1, AXIS_LENGTH);
        xAxis.setMaterial(redMaterial);
        yAxis.setMaterial(greenMaterial);
        zAxis.setMaterial(blueMaterial);
        axisGroup.getChildren().addAll(xAxis, yAxis, zAxis);
        axisGroup.setVisible(true);
        world.getChildren().addAll(axisGroup);
}
```

(5) 添加对 buildAxes( ) 方法的调用，如 Example 7-28 所示。

Example 7-28　Add Call to buildAxes( ) Method

```
buildCamera();
buildAxes();
```

(6) 在 IDE 中右击 MoleculeSampleApp 节点，编译并运行项目。打开项目窗口，然后选择"运行"选项。此时会出现一个包含三维轴的窗口，如图 7.2 所示。

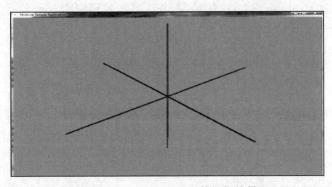

图 7.2　MoleculeSampleApp 的运行结果(1)

6. 建造分子

本节将构建 Molecular UI。这是使用 Xform 类和 3D 功能的地方，例如 PhongMaterial、Sphere 和 Cylinder，还使用了 Xform 类。

(1) 要声明分子群变换，需要复制 Example 7-29 中的代码，并将其粘贴到 axisGroup 变量之后。

Example 7-29　Declare the moleculeGroup Xform

```
final Xform axisGroup = new Xform();
final Xform moleculeGroup = new Xform();
```

(2) 为 buildMolecule( ) 方法中使用的类添加以下导入语句。

Example 7-30　Add Import Statements for build Molecule( )

```
import javafx.scene.shape.Cylinder;
import javafx.scene.shape.Sphere;
import javafx.scene.transform.Rotate;
```

(3) 添加以下 buildMolecule( ) 方法中使用的变量。

Example 7-31　Add Import Statements for buildMolecule( )

```
private static final double AXIS_LENGTH = 250.0;
private static final double HYDROGEN_ANGLE = 104.5;
```

(4) 复制 buildMolecule( ) 方法的代码，并将其粘贴到 MoleculeSampleApp 的 buildAxis( ) 方法中。

(5) 在 start( ) 方法中添加对 buildMolecule( ) 方法的调用，如 Example 7-32 所示。

Example 7-32　Add the Call to the buildMolecule( ) Method

```
buildCamera();
buildAxes();
buildMolecule();
```

(6) 运行该项目，运行结果如图 7.3 所示。

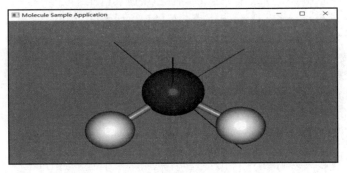

图 7.3　MoleculeSampleApp 的运行结果(2)

(7) 通过将 visible 属性修改为 false 可以关闭轴的可见性，如 Example 7-33 所示。再次运行 MolecleSampleApp，查看正在运行的应用程序，如图 7.4 所示，但不显示轴。

Example 7-33　Set visible Property to false

```
axisGroup.setVisible(false);
```

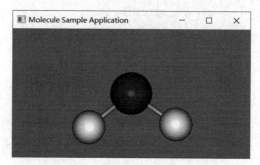

图 7.4　MoleculeSampleApp 的运行结果(3)

7. 添加照相头查看控件

handleMouse( )和 handleKeyboard( )方法允许查看不同的相机视图,用于演示如何使用鼠标和键盘在场景中操纵相机的视图。

(1) 为 handleMouse( )方法中使用的变量添加声明,并在 handleKeyboard( )方法中添加源代码。复制 Example 7-34 中的代码,并将其粘贴在 HYDROGEN_ANGLE 声明行后。

Example 7-34　Add Variables Used

```
private static final double CONTROL_MULTIPLIER = 0.1;
private static final double SHIFT_MULTIPLIER = 10.0;
private static final double MOUSE_SPEED = 0.1;
private static final double ROTATION_SPEED = 2.0;
private static final double TRACK_SPEED = 0.3;
double mousePosX;
double mousePosY;
double mouseOldX;
double mouseOldY;
double mouseDeltaX;
double mouseDeltaY;
```

(2) 复制 handleMouse( )和 handleKeyboard( )方法中使用的导入语句,如 Example 7-35 所示。将它们粘贴在 MolecleSampleApp.java 文件的顶部。

Example 7-35　Add the Import Statements

```
import javafx.event.EventHandler;
import javafx.scene.input.KeyEvent;
import javafx.scene.input.MouseEvent;
```

（3）从附录 1 的 MolecleSampleApp 代码中复制 handleMouse（ ）和 handleKeyboard（ ）方法的代码行，将它们添加到 molecleSampleApp.java 文件的 buildMolecule（ ）方法中。在 start（ ）方法中，添加对刚才添加的 handleKeyboard（ ）和 handleMouse（ ）方法的调用。复制 Example 7-36 中的代码，并将其粘贴到场景之后，用于设置填充（颜色为灰色）线。

Example 7-36　Add Method Calls

```
Scene scene = new Scene(root, 1024, 768, true);
scene.setFill(Color.GREY);
handleKeyboard(scene, world);
handleMouse(scene, world);
```

（4）保存文件。编译并运行项目，使用鼠标或键盘获得不同的视图。
- 按住鼠标左键，向右或向左上下拖曳鼠标。
- 绕轴旋转模型的相机视图。
- 按住鼠标右键并向左拖曳鼠标以移动相机。
- 远离模型查看，向右拖曳鼠标以移动相机。
- 更接近分子模型的视图。
- 按 Ctrl+Z 键将模型返回到其初始位置。
- 按 Ctrl+V 键在视图中显示和隐藏分子。
- 按 Ctrl+X 键显示和隐藏轴。

7.8　Canvas

本节探讨 JavaFX Canvas API，并提供可以编译和运行的代码示例。JavaFX Canvas API 提供了一个自定义纹理，可以对其进行写入，它由 JavaFX 中的 Canvas 类和 GraphicsContext 类定义。使用该 API 需要创建 Canvas 对象，获取其 GraphicsContext，并调用绘图操作以在屏幕上呈现自定义的形状。因为 Canvas 是一个节点的子类，所以可以在 JavaFX 场景图中使用它。

1. 绘制基本形状

BasicOpsTest 项目（运行结果如图 7.5 所示）创建了一个画布，获取其 GraphicsContext，并为其绘制一些基本形状。直线、椭圆形、圆形、矩形、圆弧和多边形都可以使用 GraphicsContext 类中定义的方法绘制。

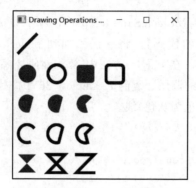

图 7.5　BasicOpsTest 的运行结果

Example 7-37　Drawing Some Basic Shapes on a Canvas

```
import javafx.application.Application;
import javafx.scene.Group;
import javafx.scene.Scene;
import javafx.scene.canvas.Canvas;
import javafx.scene.canvas.GraphicsContext;
import javafx.scene.paint.Color;
import javafx.scene.shape.ArcType;
import javafx.stage.Stage;
public class BasicOpsTest extends Application {
    public static void main(String[] args) {
      launch(args);
    }
    @Override
    public void start(Stage primaryStage) {
        primaryStage.setTitle("Drawing Operations Test");
        Group root = new Group();
        Canvas canvas = new Canvas(300, 250);
        GraphicsContext gc = canvas.getGraphicsContext2D();
        drawShapes(gc);
        root.getChildren().add(canvas);
```

```
            primaryStage.setScene(new Scene(root));
            primaryStage.show();
        }
        private void drawShapes(GraphicsContext gc) {
          gc.setFill(Color.GREEN);
          gc.setStroke(Color.BLUE);
          gc.setLineWidth(5);
          gc.strokeLine(40, 10, 10, 40);
          gc.fillOval(10, 60, 30, 30);
          gc.strokeOval(60, 60, 30, 30);
          gc.fillRoundRect(110, 60, 30, 30, 10, 10);
          gc.strokeRoundRect(160, 60, 30, 30, 10, 10);
          gc.fillArc(10, 110, 30, 30, 45, 240, ArcType.OPEN);
          gc.fillArc(60, 110, 30, 30, 45, 240, ArcType.CHORD);
          gc.fillArc(110, 110, 30, 30, 45, 240, ArcType.ROUND);
          gc.strokeArc(10, 160, 30, 30, 45, 240, ArcType.OPEN);
          gc.strokeArc(60, 160, 30, 30, 45, 240, ArcType.CHORD);
          gc.strokeArc(110, 160, 30, 30, 45, 240, ArcType.ROUND);
          gc.fillPolygon(new double[]{10, 40, 10, 40},
          new double[]{210, 210, 240, 240}, 4);
          gc.strokePolygon(new double[]{60, 90, 60, 90},
          new double[]{210, 210, 240, 240}, 4);
          gc.strokePolyline(new double[]{110, 140, 110, 140},
          new double[]{210, 210, 240, 240}, 4);
        }
    }
```

如上所示,画布的宽度为 300,高度为 250。然后调用 getGraphicsContext2D( )方法,再调用 strokeLine( )、fillOval( )、strokeac( )和 fillPolygon( )等方法以执行一些基本的绘图操作。

2. 应用渐变和阴影

下一个示例(CanvasTest 项目)通过绘制自定义形状以及一些渐变和阴影测试更多的 GraphicsContext 方法。运行结果如图 7.6 所示。

本示例的代码经过组织,每个绘图操作都是

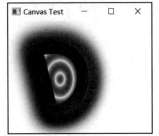

图 7.6 CanvasTest 的运行结果

其自身的私有方法,这就允许通过简单的操作测试不同的功能调用(或注释掉)感兴趣的方法。在学习 Canvas API 时,需要关注的代码是对 Canvas 或 API 的底层调用 GraphicsContext 对象。

这个模式有 5 个主要部分。首先,画布的位置设置为坐标(0,0)。这是通过调用 Example 7-38 中的代码将 Translate 转换应用于底层画布对象实现的。

Example 7-38　Moving the Canvas

```
private void moveCanvas(int x, int y) {
    canvas.setTranslateX(x);
    canvas.setTranslateY(y);
}
```

可以将其他值作为参数传入,以将画布移动到新位置。传入的值将转发到 setTranslateX 和 setTranslateY 方法中,画布将相应的移动。接下来,在屏幕上绘制主形状(看起来像大写字母 D)。这是通过 bezier 曲线完成的,可以通过调用 GraphicsContecxt 对象的 bezierCurveTo( )方法实现。

Example 7-39　Drawing a Bezier Curve（Capital "D"）On Screen

```
private void drawDShape() {
gc.beginPath();
gc.moveTo(50, 50);
gc.bezierCurveTo(150, 20, 150, 150, 75, 150);
gc.closePath();
}
```

可以通过更改参数值试验这个形状。这个 BezierCurveto 可以拉伸形状。之后,红色和黄色的径向渐变提供了背景中出现的圆形图案。

Example 7-40　Drawing a RadialGradient

```
private void drawRadialGradient(Color firstColor, Color lastColor) {
gc.setFill(new RadialGradient(0, 0, 0.5, 0.5, 0.1, true,
CycleMethod.REFLECT,
new Stop(0.0, firstColor),
new Stop(1.0, lastColor)));
gc.fill();
}
```

这里,GraphicsContext 的 setFill( )方法接受 RadialGradient 对象作为其参

数。同样,可以尝试使用不同的值,或者根据自己的喜好使用不同的颜色。

LinearGradient 将自定义 D 形从蓝色变为绿色。

Example 7-41   Drawing a LinearGradient

```
private void drawLinearGradient(Color firstColor, Color
secondColor) {
    LinearGradient lg = new LinearGradient (0, 0, 1, 1, true,
CycleMethod.REFLECT, new Stop (0.0, firstColor), new Stop (1.0,
secondColor));
    gc.setStroke(lg);
    gc.setLineWidth(20);
    gc.stroke();
}
```

这段代码将 GraphicsContext 的笔画设置为使用 LinearGradient,然后使用 gc.stroke( ) 方法呈现图案。最后,通过调用 GraphicContext 对象上的 applyEffect 提供多色阴影。

Example 7-42   Adding a DropShadow

```
private void drawDropShadow (Color firstColor, Color secondColor,
Color thirdColor, Color fourthColor) {
    gc.applyEffect(new DropShadow(20, 20, 0, firstColor));
    gc.applyEffect(new DropShadow(20, 0, 20, secondColor));
    gc.applyEffect(new DropShadow(20, -20, 0, thirdColor));
    gc.applyEffect(new DropShadow(20, 0, -20, fourthColor));
}
```

如 Example 7-42 所示,通过创建具有指定颜色的 DropShadow 对象应用该效果,颜色将传递给 GraphicsContextobject 对象的 applyEffect( )方法。

3. 与用户交互

在下面的示例(CanvasDoodleTest)中,屏幕上会出现一个蓝色方块,当用户拖曳光标穿过其表面时,该方块会被慢慢删除,如图 7.7 所示。

现在已经了解了如何创建基本形状和渐变。因此,只关注 Example 7-43 中负责与用户交互的代码。

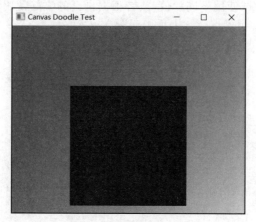

图 7.7 与用户交互(1)

Example 7-43　Interacting with the User

```
...
private void reset(Canvas canvas, Color color) {
    GraphicsContext gc = canvas.getGraphicsContext2D();
    gc.setFill(color);
    gc.fillRect(0, 0, canvas.getWidth(), canvas.getHeight());
}
@Override
public void start(Stage primaryStage) {
    ...
    final GraphicsContext gc = canvas.getGraphicsContext2D();
    ...
    //用户拖曳光标时清除部分
    canvas. addEventHandler ( MouseEvent. MOUSE _ DRAGGED, new
EventHandler<MouseEvent>() {
    @Override
    public void handle(MouseEvent e) {
        gc.clearRect(e.getX() - 2, e.getY() - 2, 5, 5);
    }
});
    //当用户双击时,用蓝色矩形填充画布
    canvas. addEventHandler ( MouseEvent. MOUSE _ CLICKED, new
EventHandler<MouseEvent>() {
    @Override
```

```
      public void handle(MouseEvent t) {
        if (t.getClickCount( ) >1) {
          reset(canvas, Color.BLUE);
        }
      }
    });
    ...
```

Example 7-43 中定义了一个重置方法,用默认的蓝色填充整个矩形。在 start()方法中,它覆盖了与用户的交互。第一个注释部分向进程添加了一个事件处理程序,当用户拖曳光标时,就会调用 MouseeEvent 对象。每次拖曳,clearRect 都会调用 GraphicsContext 对象的方法,以传入当前光标的坐标和要清除的区域的大小。当这种情况发生时,背景渐变将显示出来,如图 7.8 所示。

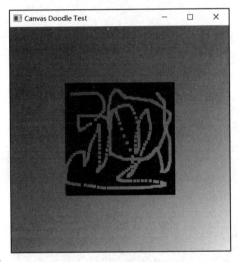

图 7.8　与用户交互(2)

剩下的代码只是计算点击次数,如果用户双击鼠标,就会将蓝色方块重置为其原始状态。

4. 创建一个简单的图层系统

可以通过实例化多个 Canvas 对象定义一个简单的图层系统。因此,切换层就成了选择所需画布的问题(画布对象是完全透明的,直到可以利用它的一部分)。最后,LayerTest 项目添加了两个画布对象,并直接放置在彼此的顶部。当点击屏幕时,一个彩色圆圈将出现在当前选定的图层上。可以使用屏幕顶部的 ChoiceBox 改变图层。添加到第 1 层的圆圈将显示为绿色，添加到第 2 层的圆圈将显示为蓝色，如图 7.9 所示。

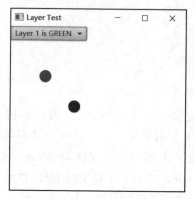

图 7.9　LayerTest.java 的执行结果(1)

该演示的 GUI 使用边框窗格管理其组件。一个 ChoiceBox 被添加到了屏幕的顶部，两个画布对象被添加到了一个面板，面板被添加到了屏幕的中心。

Example 7-44　Creating and Adding the Layers

```
...
private void createLayers() {
    //第 1 层和第 2 层的大小相同
    layer1 = new Canvas(300,250);
    layer2 = new Canvas(300,250);
    //Obtain Graphics Contexts
    gc1 = layer1.getGraphicsContext2D();
    gc1.setFill(Color.GREEN);
    gc1.fillOval(50,50,20,20);
    gc2 = layer2.getGraphicsContext2D();
    gc2.setFill(Color.BLUE);
    gc2.fillOval(100,100,20,20);
}
...
private void addLayers() {
    //添加图层
    borderPane.setTop(cb);
    Pane pane = new Pane();
    pane.getChildren().add(layer1);
```

```
        pane.getChildren().add(layer2);
        layer1.toFront();
        borderPane.setCenter(pane);
        root.getChildren().add(borderPane);
    }
    ...
```

用户交互是通过将事件处理程序直接添加到每个层完成的。点击画布将生成一个 MouseeEvent，当收到它时，将在当前光标位置绘制一个圆圈。

Example 7-45　Adding Event Handlers

```
private void handleLayers() {
    //第1层的事件处理程序
    layer1.addEventHandler(MouseEvent.MOUSE_PRESSED, new EventHandler<MouseEvent>() {
        @Override
        public void handle(MouseEvent e) {
            gc1.fillOval(e.getX(),e.getY(),20,20);
        }
    });
    //第2层的事件处理程序
    layer2.addEventHandler(MouseEvent.MOUSE_PRESSED, new EventHandler<MouseEvent>() {
        @Override
        public void handle(MouseEvent e) {
            gc2.fillOval(e.getX(),e.getY(),20,20);
        }
    });
}
```

因为这两个层都直接放置在彼此的顶部，所以只有最顶部的画布才会处理鼠标点击。要将特定层移动到堆栈的前面，只需要在屏幕顶部的 ChoiceBox 组件中选择它。

Example 7-46　Selecting a Layer

```
private void createChoiceBox() {
    cb = new ChoiceBox();
    cb.setItems(FXCollections.observableArrayList("Layer 1 is GREEN", "Layer 2 is BLUE"));
```

```
        cb.getSelectionModel().selectedItemProperty().addListener(new
    ChangeListener() {
        @Override
        public void changed(ObservableValue o, Object o1, Object o2) {
            if(o2.toString().equals("Layer 1 is GREEN")) {
                layer1.toFront();
            }
            else if(o2.toString().equals("Layer 2 is BLUE")) {
                layer2.toFront();
            }
        }
    });
    cb.setValue("Layer 1 is GREEN");
}
```

如 Example 7-46 所示,ChangleListener 在 ChoiceBox 上注册,并通过调用相应画布上的 toFront() 方法将所选层置于前景。在添加了大量蓝色和绿色圆圈后切换图层时,图层选择将变得更加明显。可以通过观察圆圈的边缘判断哪一层被移到了前面,如图 7.10 与图 7.11 所示。

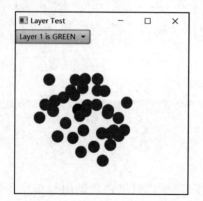 

图 7.10　LayerTest.java 的执行结果(2)　　图 7.11　LayerTest.java 的执行结果(3)

选择层的能力在软件应用程序中很常见,例如图像操纵程序。因为每个画布对象都是一个节点,所以可以自由地将所有标准转换和视觉效果应用到其他组件上。

7.9 小结

本章介绍了 JavaFX 的图形技术，包括 JavaFX 3D 图形入门、使用图像操作 API 以及使用 Canvas API。每部分都提供了有关 JavaFX 图形功能可用的 API 信息，包括代码示例和应用程序，以及如何使用这些 API；还引导读者运用所学知识构建了一个综合的示例应用程序。读者通过学习这些知识能够掌握基于 JavaFX 的 3D 图形开发技术。

第 8 章 JavaFX Web 应用开发

本章将介绍 JavaFX 嵌入式浏览器,这是一个用户界面组件,通过其 API 能够提供 Web 查看器和浏览器的功能。本章包含以下知识点:

① JavaFX WebView 组件概述——列出 WebView 组件的基本功能,并介绍 javafx.scene.web 包;

② 支持 HTML5 的功能——描述 WebView 组件支持的 HTML5 的功能;

③ 将 WebView 组件添加到 JavaFX 应用场景中——提供有关如何在 WebView 中嵌入基于浏览器的相关知识的讨论,以及如何将组件添加到 JavaFX 应用场景中;

④ 处理 JavaScript 命令——介绍如何为当前的文档运行特定的 JavaScript 命令,并加载到嵌入式浏览器中;

⑤ 从 JavaScript 调用 JavaFX 应用——提供有关如何实现从 Web 内容到 JavaFX 的调用的讨论;

⑥ 管理 Web 弹出窗口——讨论如何使用 PopupFeatures 类为其设置其他 WebView 对象,并在单独窗口中打开文档;

⑦ 管理 Web 历史记录——说明如何使用 WebHistory 类获取已访问页面的列表;

⑧ HTML 内容打印——提供用于打印嵌入式浏览器的 HTML 内容的代码模式。

8.1 JavaFX WebView 组件概述

JavaFX 嵌入式浏览器是一个用户界面组件,通过其 API 提供了 Web 查看器和浏览器的功能。嵌入式浏览器组件基于开源的 Web 浏览器引擎 WebKit,支持级联样式表(CSS)、JavaScript、文档对象模型(DOM)和 HTML5。嵌入式浏览器能够在 JavaFX 中执行以下任务:

① 从本地和远程 URL 中呈现 HTML 内容;

② 获取网络历史记录；
③ 执行 JavaScript 命令；
④ 执行从 JavaScript 到 JavaFX 的向上调用；
⑤ 管理 Web 弹出窗口；
⑥ 将效果应用于嵌入式浏览器。

嵌入式浏览器从 Node 类继承了其字段和方法，因此拥有它的所有特性。构成嵌入式浏览器的类位于 javafx.scene.web 包中。图 8.1 展示了嵌入式浏览器的体系结构及其与其他 JavaFX 类的关系。

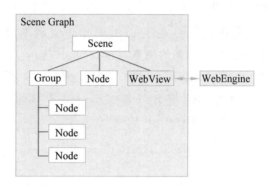

图 8.1　嵌入式浏览器的体系结构及其与其他类的关系

8.1.1　WebEngine 类

WebEngine 类提供了基本的网页功能，并支持与用户的交互，例如导航链接和提交 HTML 表单。但是，它不能实现与用户的直接交互。WebEngine 类一次处理一个网页，支持加载 HTML 内容、访问 DOM 以及执行 JavaScript 命令等基本浏览功能。

可以用两个构造方法创建 WebEngine 对象：空构造方法和具有指定 URL 值的构造方法。如果实例化空构造方法，则可以通过 load() 方法将 URL 值传递给 WebEngine 对象。启动 JavaFX SDK 2.2，开发人员可以启用或禁用特定 Web 引擎的 JavaScript 调用，并应用自定义样式表，用户样式表会将该 WebEngine 实例中呈现的页面上的默认样式替换为用户定义的样式。

8.1.2　WebView 类

WebView 类是节点类（Node）的扩展，封装了 WebEngine 对象，可以将 HTML 内容合并到应用程序的场景中，并提供应用效果和转换的属性和方法。

对 WebView 对象调用 getEngine( ) 方法,将返回与之相关联的 Web 引擎。Example 8-1 中的 WebView 和 WebEngine 展示了创建应用对象的一般方式。

Example 8-1　Creating WebView and WebEngine Objects

```
WebView browser = new WebView( );
WebEngine webEngine = browser.getEngine( );
webEngine.load("http://mySite.com");
```

8.1.3　PopupFeatures 类

PopupFeatures 类描述了 JavaScript 规范中定义的 Web 弹出式窗口的功能。当需要在应用中打开一个新的浏览器窗口时,该类的实例对象将通过调用 setCreatePopupHandler( ) 方法传递值到 WebEngine 对象上注册的弹出式窗口的处理程序中,如 Example 8-2 所示。

Example 8-2　Creating a Pop-Up Handler

```
webEngine. setCreatePopupHandler ( new  Callback < PopupFeatures,
WebEngine>( ) {
    @Override public WebEngine call(PopupFeatures config) {
        //do something
        //return a web engine for the new browser window
    }
});
```

如果该方法返回同一个 WebView 对象的 Web 引擎,则目标文档将在同一个浏览器窗口中打开。要在另一个窗口中打开目标文档,可以指定另一个 Web 视图的 WebEngine 对象。当需要阻止弹出窗口时,将返回一个空值。

8.1.4　其他特性

WebView 组件具有默认的内存缓存,这意味着一旦应用包含的 WebView 组件被关闭时,任何缓存的内容都将会丢失。但是,开发人员可以通过实现 java.net.ResponseCache 类在应用程序级别进行缓存。从 WebKit 的角度来看,持久缓存是网络层的一个属性,类似于连接 Cookie 处理程序。一旦安装了它,WebView 组件就会以透明的方式使用它们。

8.2 JavaFX 支持的 HTML5 功能

本节将介绍 JavaFX Web 组件支持的 HTML5 功能,大多数支持的功能都是 WebEngine 类和 WebView 类实现的一部分,并且该功能没有公共 API。JavaFX Web 组件的实现支持以下 HTML5 功能:
- 画布和 SVG
- 媒体播放
- 表单控件
- 历史维护
- 交互式元素标签
- DOM
- WebWorkers
- Web 套接字
- 网络字体

8.2.1 Canvas 与 SVG

对 Canvas 和 SVG(可缩放矢量图形)元素标记的支持实现了基本的图形功能,包括渲染图形,使用 SVG 的语法构建形状,设置应用的颜色,以及实现视觉效果和动画。Example 8-3 提供了一个使用＜canvas＞和＜svg＞标记呈现 Web 组件的测试。

Example 8-3　Use of Canvas and SVG Elements

```
<!DOCTYPE HTML>
<html><head><title>Canvas and SVG</title>
<canvas style="border:3px solid darkseagreen;" width="200" height=
"100"></canvas>
<svg><circle cx="100" cy="100" r="50" stroke="black" stroke-width
="2" fill="red"/></svg>
</body></html>
```

当使用 Example 8-3 中的 HTML 代码将页面加载到 WebViewSample 程序时,其执行结果如图 8.2 所示(WebViewSample.java 的源代码详见附录 2)。

8.2.2 媒体播放

WebView 组件能够在加载的 HTML 页面中播放视频和音频。WebView

图 8.2　WebViewSample 程序的执行结果(1)

组件支持以下编解码器。
- 音频：AIFF、WAV(PCM)、MP3 和 AAC。
- 视频：VP6、H264。
- 媒体容器：FLV、FXM、MP4 和 MpegTS(HLS)。

有关嵌入媒体内容的更多信息，请参阅视频和音频标签的 HTML5 规范。

8.2.3　表单控制

JavaFX Web 组件支持呈现表单以处理数据输入。支持的表单控件包括文本字段、按钮、复选框和其他可用的输入控件。Example 8-4 提供了一组控件，能够输入问题摘要并指定其优先级。

Example 8-4　Form Input Controls

```
<!DOCTYPE HTML>
<html>
    <form>
        <p><label>Login: <input></label></p>
        <fieldset>
          <legend> Priority </legend>
            <p><label> <input type=radio name=size> High </label>
</p>
```

```
                <p><label> <input type=radio name=size> Medium </label>
</p>
                <p><label> <input type=radio name=size> Low </label></p>
        </fieldset>
    </form>
</html>
```

当 Example 8-4 中的 HTML 内容加载到 WebViewSample（WebViewSample.java 的源代码详见附录 2）程序时，它会产生如图 8.3 所示的输出。

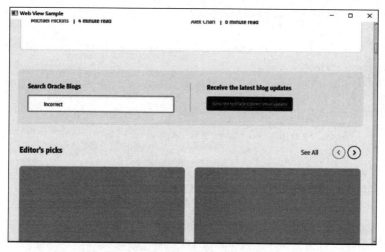

图 8.3　WebViewSample 程序的执行结果（2）

有关用户如何使用表单控件提交和处理数据的更多信息，请参阅 HTML5 规范。

8.3　历史记录维护

可以使用 JavaFX 中 javafx.scene.web 包提供的 WebHistory 类获取访问页面的列表。WebHistory 类表示与 WebEngine 对象关联的会话历史。该功能将在 WebViewSample 应用程序中启用，有兴趣的读者可以通过该应用程序了解此 JavaFX Web 组件的历史记录的维护功能。

8.4 交互式元素标记

WebView 组件支持交互式 HTML5 元素,如详细信息、摘要、命令和菜单。Example 8-5 展示了如何在 Web 组件中呈现细节和摘要元素,该示例还使用进度和仪表控制元素。

Example 8-5　Use of Details, Summary, Progress, and Meter Elements

```
<!DOCTYPE HTML>
<html><h1>Download Statistics</h1>
   <details>
       <summary>Downloading... <progress max="100" value="25"></progress> 25%</summary>
       <ul>
           <li>Size: 1,7 MB</li>
           <li>Server: oracle.com</li>
           <li>Estimated time: 2 min</li>
       </ul>
   </details>
   <h1>Hard Disk Availability</h1>
       System (C:) <meter value=240 max=326></meter> </br>
       Data (D:) <meter value=101 max=130></meter>
</html>
```

当 Example 8-5 中的 HTML 内容加载到 WebViewSample 程序时,它会产生如图 8.4 所示的输出结果(WebViewSample.java 的源代码详见附录 2)。

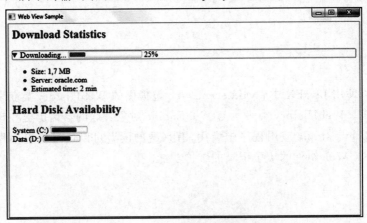

图 8.4　呈现交互式 HTML5 元素

有关交互元素属性的更多信息,请参阅 HTML5 规范。

8.5 文档对象模型

WebEngine 对象是 JavaFX Web 组件的非可视部分,可以创建并提供对网页文档对象模型(DOM)的访问。可以使用 WebEngine 类的 getDocument()方法访问文档模型的根元素。Example 8-6 提供了一个代码片段以实现一些任务——获取网页的 URI 并将其显示在应用程序窗口的标题中。

Example 8-6　Deriving a URI from a DOM

```
WebView browser = new WebView();
WebEngine webEngine = browser.getEngine();
stage.setTitle(webEngine.getDocument().getDocumentURI());
```

此外,WebEngine 对象还支持文档模型事件的规范,以定义 JavaFX 代码中的事件处理程序。有关将事件侦听器附加到网页元素的示例,有兴趣的读者请参阅 WebEngine 类的规范。

8.6 Web Sockets

WebView 组件支持 WebSocket 接口,使 JavaFX 应用程序能够与服务器进程建立双向通信。WebSocket API 规范中详细描述了 WebSocket 接口。Example 8-7 展示了使用 Web 套接字的常见模型。

Example 8-7　Using Web Sockets in HTML Code

```
<!DOCTYPE HTML>
<html><head><title>Web Sockets</title></head>
  <body>
    <script> socketConnection = new WebSocket('ws://example.com:8001');
        socketConnection.onopen = function() {
            //do some stuff
        };
    </script>
  </body>
</html>
```

8.7 Web Workers

JavaFX Web 组件支持加载的网页上的活动,并且可以并行运行 Web Worker 脚本。该功能允许执行长时间运行的脚本,而无须等待用户的交互。Example 8-8 展示了一个使用 myWorker 脚本执行长时间运行任务的网页。

Example 8-8　Using a Web Worker Script

```
<!DOCTYPE HTML>
<html>
    <head>
        <title>Web Worker</title>
    </head>
        <body>
          <script>
            var worker = new Worker('myWorker.js');
            worker.onmessage = function (event) {
                document.getElementById('result').textContent = event.data;
            };
          </script>
        </body>
</html>
```

8.8 Web Font

JavaFX Web 组件支持使用@font-face 规则声明的 Web 字体,该规则允许链接在需要时自动下载的字体。根据 HTML5 规范,该功能提供了选择与给定页面的设计目标紧密匹配的字体的功能。Example 8-9 中的 HTML 代码使用 @font-face 规则链接其 URL 指定的字体。

Example 8-9　Using a Web Font

```
<!DOCTYPE HTML>
<html><head><title>Web Font</title>
      <style>
        @font-face {
```

```
            font-family: "MyWebFont";
            src: url("http://example.com/fonts/MyWebFont.ttf")
        }
        h1 { font-family: "MyWebFont", serif;}
    </style>
  </head><body><h1> This is a H1 heading styled with MyWebFont</h1>
</body>
</html>
```

当将这个 HTML 代码加载到 WebViewSample 应用程序时(WebViewSample.java 的源代码详见附录 2),其执行结果如图 8.5 所示。

图 8.5　WebViewSample 程序的执行结果(3)

8.9　将 WebView 组件添加到应用场景中

本节将介绍 WebViewSample 应用程序,讨论如何实现将 WebView 组件添加到 JavaFX 场景中的任务,并应用 WebViewSample 应用程序创建一个 Browser 类,该类用 UI 控件封装 WebView 对象和工具栏。应用程序的 WebViewSample 类将创建场景,并将浏览器对象添加到场景中。以下示例展示了如何将 WebView 组件添加到应用程序场景中(WebViewSample.java 的源代码详见附录 2)。

在这段代码中,Web 引擎加载指向 Oracle 公司网站的 URL,包含该 Web 引擎的 WebView 对象通过使用 getChildren()和 add()方法添加到应用程序场景中。当添加、编译并运行此代码片段时,它会生成如图 8.6 所示的应用程序窗口。

图 8.6　WebViewSample 程序的执行结果（4）

8.10　创建工具栏

本节将添加带有 4 个超链接对象的工具栏，以便在不同的 Oracle Web 资源之间切换。学习示例所示的 Browser 类的代码（WebViewSample.java 的源代码详见附录 2），它为其他 Web 资源添加 URL，包括 Oracle 产品、博客、Java 文档和合作伙伴网络。代码片段还创建了一个工具栏，并向其中添加了超文本链接。

这段代码使用 for 循环创建超文本链接。setOnAction()方法定义超链接的行为。当用户单击链接时，相应的 URL 值会传递给 WebEngine 类的 load()方法。编译并运行修改后的应用程序时，应用程序窗口会发生变化，如图 8.7 所示。

图 8.7　WebViewSample 程序的执行结果（5）

8.11 调用 JavaScript 命令

本节将扩展 WebViewSample 应用程序,并讨论如何用 JavaFX 代码调用 JavaScript 命令。WebEngine 类提供了在当前 HTML 页面的上下文环境中运行脚本的 API 的功能。

1. executeScript()方法

WebEngine 类的 executeScript()方法允许执行任何用 JavaScript 在加载的 HTML 页面中声明的命令。可以使用以下字符串调用此 Web 引擎上的方法:

```
webEngine.executeScript("<function name>");
```

该方法的执行结果被转换为 java.lang,通过以下规则即可进行转换:
- JavaScript Int 32 被转换为 Java 整型;
- JavaScript 数字被转换为 java.Double;
- JavaScript 字符串值被转换为 java.String;
- JavaScript 布尔值被转换为 java.Boolean。

有关 WebEngine 类的更多信息,请参阅 API 文档的转换结果。

2. 从 JavaFX 代码调用 JavaScript 命令

扩展 WebViewSample 应用程序以引入"帮助"文件并执行 JavaScript 命令,用于切换"帮助"文件中的主题列表。创建指向帮助的"帮助"工具栏项,利用 HTML 文件添加帮助,用户可以在其中预览有关 Oracle 网站的参考资料。将 Example 8-10 中的 HTML 文件加载到 WebViewSample 应用程序中,执行结果如图 8.7 所示。

Example 8-10　help.html file

```
<html lang="en"><head>
<!-- Visibility toggle script -->
<script type="text/javascript">
<!-- function toggle_visibility(id) {
    var e = document.getElementById(id);
    if (e.style.display == 'block')
      e.style.display = 'none';
    else
      e.style.display = 'block';
```

```
         }
//-->
</script></head><body><h1>Online Help</h1>
<p class="boxtitle"><a href="#" onclick="toggle_visibility('help_
topics');"
class="boxtitle">[+] Show/Hide Help Topics</a></p>
<ul id="help_topics" style='display:none;'>
<li> Products - Extensive overview of Oracle hardware and software
products, and summary Oracle consulting, support, and educational
services. </li>
<li>Blogs - Oracle blogging community (use the Hide All and Show All
buttons to collapse and expand the list of topics).</li>
<li> Documentation - Landing page to start learning Java. The page
contains links to the Java tutorials, developer guides, and API
documentation.</li>
< li > Partners - Oracle partner solutions and programs. Popular
resources and membership opportunities.</li>
</ul>
</body>
</html>
```

以下示例中修改后的应用程序代码创建了"帮助"工具栏项,以及一个用于隐藏和显示帮助主题的附加按钮(WebViewSample.java 的源代码详见附录2)。只有在选择"帮助"页面时,该按钮才会添加到工具栏。加载是在后台线程上进行的,启动加载的方法在安排后台作业后将立即返回。getLoadWorker()方法提供了一个 Worker 接口的实例以跟踪加载进度。如果"帮助"页面的进度状态成功,则"切换帮助主题"按钮将添加到工具栏中,如图 8.8 所示。

图 8.8　WebViewSample 程序的执行结果(6)

Example 8-11 中的 setOnAction()方法定义了"切换帮助主题"按钮的

行为。

Example 8-11　Executing a JavaScript Command

```
//set action for the button
toggleHelpTopics.setOnAction((ActionEvent t) -> {
    webEngine.executeScript("toggle_visibility('help_topics')");
});
```

当用户单击"切换帮助主题"按钮时,executeScript()方法将会为"帮助"运行"切换可见性"JavaScript 函数以切换到 HTML 页面,并显示"帮助"主题,如图 8.9 所示。当用户再次单击时,"切换可见性"功能会隐藏主题列表。

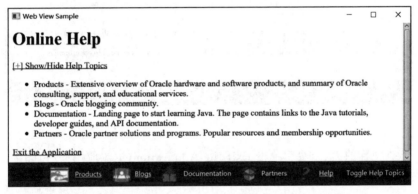

图 8.9　Web View Sample 程序的执行结果(7)

8.12　从 JavaScript 调用 JavaFX

现在知道了如何从 JavaFX 调用 JavaScript,本节将讨论相反的功能——从 JavaScript 调用 JavaFX。一般的概念是在 JavaFX 应用程序中创建一个接口对象,并通过调用 JSObject 让 JavaScript 知道它的 setMember( )方法,然后就可以调用公共方法,并从 JavaScript 访问该对象的公共字段。

首先,在帮助的 HTML 文件中再添加一行:

```
<p><a href="about:blank" onclick="app.exit()">Exit the Application
</a></p>
```

这样就可以退出应用程序。通过单击 HTML 文件的 Exit the Application 按钮中的应用程序链接,用户就可以退出 WebViewSample 应用程序。修改应用程序,以实现这个功能(WebView Sample.java 的源代码详见附录2)。

JavaApp 接口的 exit()方法是公共的,因此它可以从外部访问。调用该方法时,将会导致 JavaFX 应用程序终止运行。

WebView Sample.java 源代码中的 JavaApp 接口被设置为 JSObject 实例对象的成员,它的唯一方法可以作为应用程序退出。编译和运行 WebViewSample 应用程序,单击出现在页面底部的链接以退出应用程序,如图 8.10 所示。

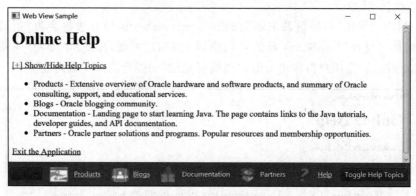

图 8.10 退出应用程序

检查文件的内容,然后单击 Exit the Application(退出应用程序)链接,就可以关闭 WebViewSample 应用程序。

8.13 管理 Web 弹出窗口

本节将介绍如何使用 WebView 组件创建浏览器中的弹出窗口,以及如何在 WebViewSample 应用程序中实现该功能。当需要在应用程序中打开新的浏览器窗口时,PopupFeatures 类的对象实例将通过 setCreatePopupHandler( )方法传递到 WebEngine 对象注册的弹出处理程序中。

在 WebViewSample 应用程序中,可以为将在单独窗口中打开的文档设置替代的 WebView 对象。图 8.11 展示了通过右击任何链接打开的上下文菜单。

要为目标文档指定新的浏览器窗口,可以使用 PopupFeatures 对象实例,如 Example 8-16 中修改的应用程序代码所示(WebViewSample.java 的源代码详见附录2)。当用户在上下文菜单中选择"在新窗口中打开链接"选项时,smallView 浏览器将添加到应用程序工具栏中。该行为由 setCreatePopupHandler()方法定义,该方法返回替代浏览器的 Web 引擎,以通知应用程序在何处呈现目标页面。编译

第 8 章　JavaFX Web 应用开发　149

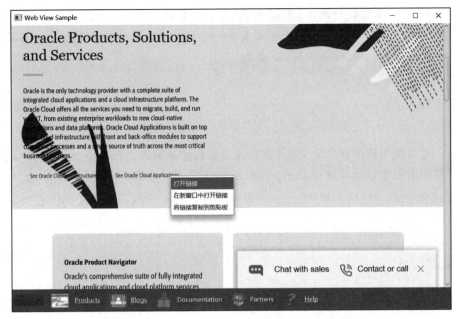

图 8.11　上下文菜单

和运行修改后的应用程序的结果如图 8.12 所示。

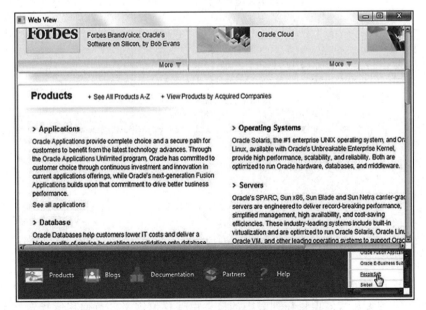

图 8.12　Web View Sample 应用程序的运行结果(8)

> 注意：在默认情况下，所有 WebView 对象都会启用上下文菜单，要禁用特定 WebView 对象实例的上下文菜单，可以将 false 值传递给 Browser 类的 setContextMenuEnabled(false) 方法。

8.14 获取访问页面列表

通常使用标准或自定义 UI 控件显示历史记录列表。Example 8-12 展示了如何获取历史记录项并在 ComboBoxcontrol 中显示它们。

Example 8-12　Obtaining and Processing the List of Web History Items

```
final WebHistory history = webEngine.getHistory();
history. getEntries ( ). addListener ( new ListChangeListener <
WebHistory.Entry>() {
    @Override
        public void onChanged(Change<? extends Entry> c) {
            c.next();
            for (Entry e : c.getRemoved()) {
              comboBox.getItems().remove(e.getUrl());     }
            for (Entry e : c.getAddedSubList()) {
              comboBox.getItems().add(e.getUrl());        }
        }
    }
);
comboBox.setPrefWidth(60);
comboBox.setOnAction(new EventHandler<ActionEvent>() {
    @Override
      public void handle(ActionEvent ev) {
        int offset = comboBox.getSelectionModel().getSelectedIndex(
) - history.getCurrentIndex();
        history.go(offset);
      }
    }
);
```

ListChangeListener 对象跟踪历史记录条目的更改，并将相应的 URL 添加到组合框中。

当用户在组合框中选择任何项目时，Web 引擎将被导航到由历史记录条目项目定义的 URL，该项目在列表中的位置由偏移值定义。负偏移值指定当前条目之前的位置，正偏移值指定当前条目之后的位置。修改后的应用程序的完整代码 WebViewSample.java 详见附录 2。编译和运行应用程序,会生成如图 8.13 所示的窗口。

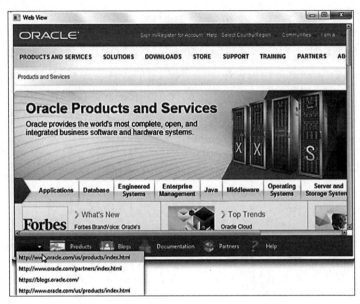

图 8.13 WebViewSample 应用程序的运行结果(9)

8.15 HTML 内容打印

本节介绍如何打印 WebView 组件中加载的网页。使用 JavaFX 8 中提供的打印 API 可以打印 JavaFX 应用程序的图形内容。打印包相应的类和枚举位于 javafx.print 包中。

8.15.1 使用打印 API

要在 JavaFX 应用程序中启用打印功能,必须使用 PrinterJob 类,该类表示与默认打印机作业关联的打印机作业系统,使用 Printer 类可以为特定作业更改打印机。当打印作业时,可以使用 JobSettings 类的属性指定作业设置,例如排序、副本、页面布局、页面范围、纸张来源、打印颜色、打印分辨率、打印质量和打印量。可以打印场景图的任何节点,包括根节点,也可以打印未添加到场景中的节点。对于特定节点,可以使用 printPage()方法启动打印作业。作业与打印页面(节点)的有关详细信息请参阅 JavaFX 8 API 规范中有关打印功能的信息。使用 JavaFX Web 组件时,通常需要打印 HTML 页面并加载到浏览器中,而不是应用程序 UI 本身,这就是要将 print()方法添加到 WebEngine 类中的原因,这种方法适用于打印与 Web 引擎关联的 HTML 页面。

8.15.2 添加上下文菜单以启用打印

通常需要向应用程序菜单添加打印命令或将打印指定给其中一个工具栏按钮。在 WebViewSample 应用程序中,工具栏上的控件会超载,这就是要将"打印"命令添加到右击启用的上下文菜单中的原因。Example 8-13 展示了一个代码片段,该代码片段将带有打印命令的上下文菜单添加到了应用程序的工具栏上。

Example 8-13　Creating a Toolbar Context Menu

```
//添加上下文菜单
final ContextMenu cm = new ContextMenu();
MenuItem cmItem1 = new MenuItem("Print");
cm.getItems().add(cmItem1);
toolBar.addEventHandler(MouseEvent.MOUSE_CLICKED, (MouseEvent e) -> {
  if (e.getButton() == MouseButton.SECONDARY) {
    cm.show(toolBar, e.getScreenX(), e.getScreenY());
  }
});
```

将 Example 8-13 中的代码片段添加到 WebViewSample 应用程序并运行,右击工具栏就会出现"打印"上下文菜单,如图 8.14 所示。

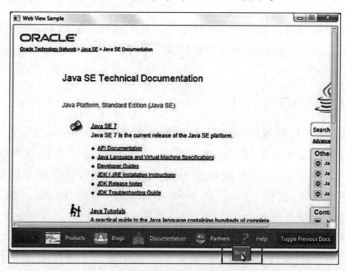

图 8.14　WebViewsample 程序的执行结果(10)

8.16 处理打印作业

将"打印"上下文菜单添加到应用程序 UI 后,可以定义印刷活动。首先,必须创建 PrinterJob 对象,然后调用 WebEngine.print 方法,将打印机作业作为参数传递的打印方法,其实现代码如 Example 8-14 所示。

Example 8-14   Calling the WebEngine.print Method

```
//处理打印作业
cmItem1.setOnAction((ActionEvent e) -> {
    PrinterJob job = PrinterJob.createPrinterJob();
    if (job != null) {
      webEngine.print(job);
      job.endJob();
    }
});
```

检查非空打印机作业很重要,因为如果系统中没有可用的打印机,则 createPrinterJob()方法将返回 Null。启用打印功能的 WebViewSample 应用程序的完整代码(详见附录 2)。

要扩展 WebViewSample 应用程序的打印功能,可以使用 javaf.print 包中提供的类。在 JavaFX 应用程序中,可以使用 TabPane 类实现浏览器选项卡,并在用户添加新选项卡时创建新的 WebView 对象。要进一步增强该应用程序的功能,可以应用效果、转换和动画过渡,还可以向应用程序场景添加更多的 WebView 实例。

有关可用功能的更多信息,请参阅 JavaFX API 文档和 JavaFX CSS 规范,还可以学习 Swing 中的 JavaFX 教程,了解如何在现有的 Swing 应用程序中添加 WebView 组件。

8.17   小结

本章介绍了 JavaFX 嵌入式浏览器这个用户界面组件,根据其 API 能够提供 Web 查看器和浏览器的功能,介绍了 JavaFX WebView 组件以及它支持的 HTML5 功能;介绍了如何将 WebView 组件添加到 JavaFX 应用程序的场景中;如何为当前文档运行特定的JavaScript命令,并将其加载到嵌入式浏览器中;如何从JavaScript调用JavaFX应用程序;讨论了如何使用PopupFeatures类为其设置其他WebView对象,并在单独窗口中打开文档;说明了如何使用WebHistory类获取已访问页面的列表;讨论了打印嵌入式浏览器的HTML内容的代码模式。

第 9 章 基于 Swing 组件的 JavaFX 应用开发

本章将介绍基于 Swing 组件的 JavaFX 应用开发，探讨如何将 JavaFX 内容集成到 Swing 应用中，以及如何在 JavaFX 应用中使用 Swing 组件，并通过若干综合性的示例介绍如何基于 Swing 组件进行 JavaFX 应用的开发。

9.1 JavaFX-Swing 的互操作性

本节将概述 JavaFX 对 GUI 开发人员的优势，阐述 JavaFX-Swing 的互操作性，展示如何丰富现有的 Swing 应用程序，以及如何实现在 JavaFX 中融合典型的 Swing 应用程序。JavaFX 对 Swing 开发人员具有明显的技术优势，旨在为应用程序提供复杂的 GUI，例如平滑动画、网络视图、音频和视频播放以及基于 CSS 等。十多年来，应用程序开发人员发现 Swing 是构建 GUI 的一个重要工具，同时是将交互性添加到 Java 应用程序的一个有效的工具包。然而，当今流行的一些 GUI 的功能却无法通过使用 Swing 组件实现。而 JavaFX 自身的优势可以帮助应用开发人员满足各种现代的需求。本章的后续内容将阐述如何使 Swing 和 JavaFX 相融合，从而开发出功能更加强大的 GUI 应用。

1. 使用 FXML

FXML 是一种基于 XML 的标记语言，它使开发人员能够在 JavaFX 应用中创建 UI，而不是实现应用程序逻辑。Swing 没有提供一种声明式方法以构建 UI，而创建 UI 的声明式方法特别适合场景图，这是因为场景图在 FXML 中更加透明。FXML 使开发人员能够更加轻松地维护复杂的 UI。

2. JavaFX 场景生成器

为了帮助开发人员构建应用程序的布局，JavaFX 提供了一个名为 JavaFX 场景生成器的设计工具。将 UI 组件拖曳到 JavaFX 的内容窗格中，该工具将生成 FXML 代码,可在 NetBeans 或 Eclipse 等 IDE 中使用这个设计工具。

3. 支持 CSS

CSS 包含控制 UI 元素外观的样式定义。在 JavaFX 应用中使用 CSS 与在 HTML 中使用 CSS 相类似。使用 CSS 可以轻松地自定义与开发 JavaFX 控件和场景图对象的主题。使用 CSS 可以将应用程序的逻辑与其视觉外观设置相分离,同时,使用 CSS 还简化了应用程序外观的进一步维护,并且能够提供一些性能上的优势。

4. 支持 JavaFX Media

借助 JavaFX 平台提供的媒体支持,通过添加音频和视频文件播放等多媒体的功能可以利用桌面应用程序,并在支持 JavaFX 的所有平台上使用媒体功能。支持的媒体编解码器的列表请参阅 JavaFX 媒体简介。更多详细的信息请参阅本书第 6 章中的 JavaFX Media 应用开发的相关内容。

5. 支持动画

动画为应用程序的界面带来了动态和现代感。在 Swing 应用中设置对象动画虽然是可能的,但其实现比较复杂。相比之下,JavaFX 使开发人员在应用中设置图形对象的动画应用变得更加容易,这是因为平台专门为此目的创建了特定的 API。

6. 支持 HTML

长期以来,Swing 开发人员一直希望能够在 Java 应用程序中呈现 HTML 内容。JavaFX 通过提供具有 Web 视图和完整浏览功能的用户界面组件实现了这一功能。有关的详细信息请参阅本书第 8 章中的基于 Web 的 JavaFX 应用开发的相关内容。

9.2 将 JavaFX 集成到 Swing 应用中

本节将介绍如何把 JavaFX 内容添加到 Swing 应用中,以及当 Swing 和 JavaFX 内容都在单个应用中运行时,如何正确地使用线程。JavaFX SDK 提供了 JFXPanel 类,这个类位于 javafx.embed.swing 软件包中,它能够将 JavaFX 内容添加到 Swing 应用中。

9.2.1 向 Swing 组件添加 JavaFX 内容

本节将创建 JFrame 组件,并添加 JFXPanel 对象,设置包含 JavaFX 的 JFXPanel 组件的图形场景。与任何 Swing 应用程序一样,可以在事件上创建

第 9 章　基于 Swing 组件的 JavaFX 应用开发

GUI 调度线程 EDT(事件调度线程)。如 Example 9-1 所示的 initAndShowGUI() 方法,该方法创建了一个 JFrame 组件,并向其中添加了一个 JFXPanel 对象,创建了 JFXPanel 类的对象并隐式启动了 JavaFX。创建 GUI 后,调用 initFX() 方法在 JavaFX 应用程序线程上创建了 JavaFX 场景。

Example 9-1

```
import javafx.application.Platform;
import javafx.embed.swing.JFXPanel;
import javafx.scene.Group;
import javafx.scene.Scene;
import javafx.scene.paint.Color;
import javafx.scene.text.Font;
import javafx.scene.text.Text;
import javax.swing.JFrame;
import javax.swing.SwingUtilities;
public class Test {
    private static void initAndShowGUI() {
        //此方法在 EDT 线程上调用
        JFrame frame = new JFrame("Swing and JavaFX");
        final JFXPanel fxPanel = new JFXPanel();
        frame.add(fxPanel);
        frame.setSize(300, 200);
        frame.setVisible(true);
        frame.setDefaultCloseOperation(JFrame.EXIT_ON_CLOSE);
        Platform.runLater(new Runnable() {
            @Override
            public void run() {
                initFX(fxPanel);
            }
        });
    }
    private static void initFX(JFXPanel fxPanel) {
        //此方法在 JavaFX 线程上调用
        Scene scene = createScene();
        fxPanel.setScene(scene);
    }
    private static Scene createScene() {
        Group root = new Group();
```

```
            Scene scene = new Scene(root, Color.ALICEBLUE);
            Text text = new Text();
            text.setX(40);
            text.setY(100);
            text.setFont(new Font(25));
            text.setText("Welcome JavaFX!");
            root.getChildren().add(text);
            return (scene);
        }
        public static void main(String[] args) {
            SwingUtilities.invokeLater(new Runnable() {
                @Override
                public void run() {
                    initAndShowGUI();
                }
            });
        }
    }
```

NetBeans IDE 13.0 版本为带有嵌入式 JavaFX 内容的 Swing 应用程序提供了支持,这样在创建 JavaFX 新项目时,就可以在 JavaFX 类别中选择 JavaFX in Swing Application。在 NetBeans IDE 中创建这个 JavaFX 应用并运行它,其运行结果如图 9.1 所示。

图 9.1 Test.java 的运行结果

9.2.2 Swing-JavaFX 的互操作性与线程

由于 JavaFX 和 Swing 数据共存于单个应用程序中,所以可能会遇到以下互操作的情况。

 • JavaFX 数据的更改由 Swing 数据所引发的更改触发。

- Swing 数据的更改由 JavaFX 数据所引发的更改触发。

1. 更改 JavaFX 数据以响应 Swing 数据的更改

由于只能在 JavaFX 用户线程上访问 JavaFX 数据,所以每当必须更改 JavaFX 数据时,就需要将代码包装成可运行的对象,并调用 Platform.runLater( ) 方法,如 Example 9-2 所示。

Example 9-2

```
jbutton.addActionListener(new ActionListener( ) {
    public void actionPerformed(ActionEvent e) {
        Platform.runLater(new Runnable( ) {
            @Override
            public void run( ) { fxlabel.setText("Swing button clicked!"); }
        });
    }
});
```

2. 更改 Swing 数据以响应 JavaFX 数据的更改

由于只能在 EDT 上更改 Swing 数据,并确保代码在 EDT 上实现,所以需要将其包装为可运行的对象,并调用 SwingUtilities.invokeLater()方法实现,如 Example 9-3 所示。

Example 9-3

```
SwingUtilities.invokeLater(new Runnable( ) {
    @Override
    public void run( ) { //Code to change Swing data. }
});
```

9.3　SimpleSwingBrowser 应用

本节介绍 Swing JavaFX 的互操作是如何工作的,并给出一个 JavaFX 综合应用的开发示例——SimpleSwingBroswer.java。这是一个 Swing 应用程序,带有一个集成的 JavaFX 组件,具有查看网页的功能。可以在地址栏中输入 URL,并在程序窗口中查看加载的页面。该程序的执行结果如图 9.2 所示(注意:该应用的运行结果是动态的,因为出版社网站的页面是经常更新的)。

图 9.2 SimpleSwingBrowser.java 的执行结果

1. 初始化 Swing Data

启动 NetBeans IDE 13.0，创建 JavaFX 新项目，在 JavaFX 类别中选择 JavaFX in Swing Application。创建 SimpleSwingBrowser 应用程序，该程序的 GUI 在应用启动时将在 EDT 上创建。主要方法的实现如 Example 9-4 所示。

Example 9-4

```
public static void main(String[] args) {
    SwingUtilities.invokeLater(new Runnable() {
        @Override
        public void run() {
            SimpleSwingBrowser browser = new SimpleSwingBrowser();
            browser.setVisible(true);
            browser.loadURL("http://www.tup.com.cn");
        }
    });
}
```

SimpleSwingBrowser 类初始化 Swing 对象，并调用 initComponents()方法创建 GUI，如 Example 9-5 所示。

Example 9-5

```java
public class SimpleSwingBrowser extends JFrame {
    private final JFXPanel jfxPanel = new JFXPanel();
    private WebEngine engine;
    private final JPanel panel = new JPanel(new BorderLayout());
    private final JLabel lblStatus = new JLabel();
    private final JButton btnGo = new JButton("Go");
    private final JTextField txtURL = new JTextField();
    private final JProgressBar progressBar = new JProgressBar();
    public SimpleSwingBrowser() {
        super();
        initComponents();
    }
    private void initComponents() {
        createScene();
        ActionListener al = new ActionListener() {
            @Override
            public void actionPerformed(ActionEvent e) {
                loadURL(txtURL.getText());
            }
        };
        btnGo.addActionListener(al);
        txtURL.addActionListener(al);
        progressBar.setPreferredSize(new Dimension(150, 18));
        progressBar.setStringPainted(true);
        JPanel topBar = new JPanel(new BorderLayout(5, 0));
        topBar.setBorder(BorderFactory.createEmptyBorder(3, 5, 3, 5));
        topBar.add(txtURL, BorderLayout.CENTER);
        topBar.add(btnGo, BorderLayout.EAST);
        JPanel statusBar = new JPanel(new BorderLayout(5, 0));
        statusBar.setBorder(BorderFactory.createEmptyBorder(3, 5, 3, 5));
        statusBar.add(lblStatus, BorderLayout.CENTER);
        statusBar.add(progressBar, BorderLayout.EAST);
        panel.add(topBar, BorderLayout.NORTH);
        panel.add(jfxPanel, BorderLayout.CENTER);
        panel.add(statusBar, BorderLayout.SOUTH);
        getContentPane().add(panel);
```

```
            setPreferredSize(new Dimension(1024, 600));
            setDefaultCloseOperation(JFrame.EXIT_ON_CLOSE);
            pack();
    }
}
```

该应用程序最顶部的窗口是一个 JFrame 对象,它包含各种 Swing 组件,例如一个文本字段、一个按钮、一个进度条和一个用于显示 JavaFX 内容的 JFX 面板。

2. 加载 JavaFX 内容

该程序第一次运行时,位于 http://www.tup.com.cn 的网页将被加载到 WebView 对象中。当输入一个新的 URL 时,附加到 initComponents()方法中的 txtURL 文本字段的 action listener 将启动页面加载,如 Example 9-6 所示。

Example 9-6

```
ActionListener al = new ActionListener() {
    @Override public void actionPerformed(ActionEvent e) {
        loadURL(txtURL.getText());
    }
};
```

只能在 JavaFX 应用程序线程上访问 JavaFX 数据。loadURL()方法将代码包装成可运行的对象,并调用 Platform.runLater()方法,如 Example 9-7 所示。

Example 9-7

```
public void loadURL(final String url) {
    Platform.runLater(new Runnable() {
        @Override public void run() {
            String tmp = toURL(url);
            if (url == null) {
                tmp = toURL("http://" + url);
            }
            engine.load(tmp);
        }
    });
}
```

```
    private static String toURL(String str) {
        try {
            return new URL(str).toExternalForm();
        } catch (MalformedURLException exception) { return null; }
    }
```

3. 更新 Swing 数据

当一个新页面加载到 WebView 组件中时，将从 JavaFX 数据中检索该页面的标题，并将其传递给 Swing GUI，作为标题放置在应用程序窗口中。该行为在 createScene( ) 方法中实现，如 Example 9-8 所示。

Example 9-8

```
    private void createScene() {
        Platform.runLater(new Runnable() {
            @Override
            public void run() {
                WebView view = new WebView();
                engine = view.getEngine();
                engine.titleProperty().addListener(new ChangeListener<String>() {
                    @Override
                    public void changed(ObservableValue<? extends String> observable,
                            String oldValue, final String newValue) {
                        SwingUtilities.invokeLater(new Runnable() {
                            @Override
                            public void run() {
                                SimpleSwingBrowser.this.setTitle(newValue);
                            }
                        });
                    }
                });
            }
        });
    }
```

9.4 在 JavaFX 中实现一个 Swing 应用

本节将创建一个 Swing 应用程序，并介绍如何在 JavaFX 应用中实现它。本节的 JavaFX 应用的运行结果如图 9.3 所示，它是一个转换器应用程序，其功能是在公制和美制单位之间转换距离测量值。可以按住鼠标左键拖曳滚动条，两个文本框中的数值会随着滚动条的移动而发生相应的改变。

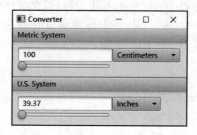

图 9.3　转换器应用程序的执行结果

1. Swing 中转换器的应用分析

转换器应用程序由以下文件组成：

- ConversionPanel.java——包含一个自定义的 JPanel 子类，用来保存组件；
- Converter.java——包含主应用程序类；
- ConverterageModel.java——定义顶部滑块的模型；
- FollowerRangeModel.java——定义底部滑块的模型；
- Units.java——创建 Unit 对象。

> 注意：每个文本字段及其滑块之间的同步均由侦听值更改的事件处理程序实现。

2. 在 JavaFX 中规划转换器应用程序

转换器应用程序包含两个类似的面板，其中包含文本字段、滑块和组合框等组件。下面将实现 ConversionPanel 类，并将该类的两个实例对象添加到转换器应用程序的图形场景中。

首先，单个 ConversionPanel 对象中的组件应当按照如下方式同步。无论何时移动滑块上的旋钮，都必须更新文本字段中的值，反之亦然；无论何时更改文本字段中的值，都必须调整滑块上旋钮的位置。一旦从组合框中选择另一个值,就必须更新文本字段的值,从而更新滑块上旋钮的位置。

其次,两个 ConversionPanel 对象都应该同步。一旦一个面板上发生更改,另一个面板上的相应组件就必须更新。建议使用 DoubleProperty 对象(称为 meters)在面板之间实现同步,并通过创建和注册两个 InvalizationListener 对象(fromMeters 和 toMeters)侦听文本字段和组合框属性的更改。每当一个面板上的文本字段的属性发生更改时,就会调用附加的 InvalidateListener 对象的 invalidated() 方法,该方法会更新 meters 属性。由于 meters 属性发生更改,因此将调用附加的 meters 属性的 InvalidateListener 对象的 invalidated() 方法,该方法将更新另一个面板上相应的文本字段。类似地,只要一个面板上的组合框的属性发生更改,就会调用附加的 InvalidateListener 对象的 invalidated() 方法,该方法会更新该面板上的文本字段。要在滑块的值和 meters 对象的值之间实现同步,就要使用双向绑定。

3. 在 JavaFX 中创建 Converter Application

在 NetBeans IDE 中创建一个新的 JavaFX 项目,并将其命名为 Converter。向该项目添加一个新的 Java 类,并将其命名为 ConversionPanel.java。

4. 创建 GUI 的标准 JavaFX 模式

在 JavaFX 中创建转换器应用程序的 GUI 之前,请参阅 Swing 应用程序中创建 GUI 的标准模式,如 Example 9-9 所示。

Example 9-9

```
public class Converter {
    private void initAndShowGUI() {
        ...
    }
    public static void main(String[] args) {
      SwingUtilities.invokeLater(new Runnable() {
        @Override
        public void run() {
          initAndShowGUI();
        }
      });
    }
}
```

要将此模式映射到 JavaFX,需要扩展 JavaFX 应用类,重写 start() 方法,并调用 main() 方法,如 Example 9-10 所示。

Example 9-10

```
import javafx.application.Application;
import javafx.stage.Stage;
public class Converter extends Application {
    @Override
    public void start(Stage t) {
        ...
    }
    public static void main(String[ ] args) {
      launch(args);
    }
}
```

当在 NetBeans IDE 中创建一个新的 JavaFX 项目时,这个模式是自动生成的。然而,了解在 JavaFX 中创建 GUI 的基本方法非常重要,尤其是在使用文本编辑器时。

5. 容器和布局

在 Swing 中,容器和布局管理器是不同的实体。创建一个容器,例如 JPanel 与 JComponent 对象,并为该容器设置布局管理器。可以用 write.add( ) 方法指定特定的布局管理器,或者不分配任何布局管理器。

在 JavaFX 中,容器本身负责布局其子节点。创造一个特定布局窗格,例如 Vbox、FlowPane 或 TitledPane 等对象,然后使用 getChildren( ).add( ) 方法将内容添加到其子节点的列表中。

JavaFX 中有几个布局容器类,称为窗格,其中一些具有它们在 Swing 中的对应项,例如 JavaFX 中的 FlowPane 类和 FlowLayout 类。

6. UI 控件

JavaFX SDK 提供了一组标准的 UI 控件。一些 UI 控件在 Swing 中有对应的控件,例如 JavaFX 中的 Button 类和 Swing 中的 JButton;JavaFX 中的滑块和 Swing 中的 JSlider;JavaFX 中的 TextField 和 Swing 中的 JTextField。要在 JavaFX 应用中实现转换器应用程序,可以使用 TextField、Slider 和 ComboBox 类提供的标准 UI 控件。

7. 获取用户操作和绑定通知的机制

在 Swing 中,可以在任何组件上注册侦听器,并侦听组件的属性,例如大小、位置或者可见性,或者监听一些事件,例如组件是否获得或失去了键盘焦点;

鼠标是否在组件上被单击、按下或释放。在 JavaFX 中，每个对象都有一组属性，可以为这些属性注册一个侦听器(listener)。只要这个属性的值发生更改，就会调用 listener。

> 注意：一个对象可以注册为另一个对象属性更改的侦听器，因此可以使用绑定机制同步两个对象的某些属性。

8. 创建 ConversionPanel 类

ConversionPanel 类用于保存组件——文本字段、滑块和组合框。创建转换器应用程序的图形场景时，需要将 ConversionPanel 类的两个实例对象添加到图形场景中。为 TitledPane 类添加导入语句，并扩展 ConversionPanel 类，如 Example 9-11 所示。

Example 9-11

```
import javafx.scene.control.TitledPane;
public class ConversionPanel extends TitledPane {
}
```

下面为 UI 控件创建实例变量。为 TextField、Slider、ComboBox 控件和 define 添加导入组件的实例变量的语句，如 Example 9-12 所示。

Example 9-12

```
import java.text.NumberFormat;
import javafx.scene.control.ComboBox;
import javafx.scene.control.Slider;
import javafx.scene.control.TextField;
private ComboBox<Unit> comboBox;
private Slider slider;
private TextField textField;
```

1) 创建 DoubleProperty 和 NumberFormat 对象

为 DoubleProperty 类和 NumberFormat 类添加 import 语句，并创建一个名为 meters 的 DoubleProperty 对象，如 Example 9-13 所示。meters 对象用于确保两个 ConversionPanel 对象之间的同步。

Example 9-13

```
import javafx.beans.property.DoubleProperty;
```

```
private DoubleProperty meters;
private numberFormat;
```

2) 布置组件

要布局文本字段和滑块,可以使用 VBox 类。要布局这两个组件和组合框,可以使用 HBox 类。为 ObservableList 类添加导入语句,并实现 ConversionPanel 类的构造方法,如 Example 9-14 所示。

Example 9-14

```
import javafx.collections.ObservableList;
public ConversionPanel (String title, ObservableList<Unit> units,
DoubleProperty meters) {
    setText(title);
    setCollapsible(false);
    numberFormat = NumberFormat.getNumberInstance();
    numberFormat.setMaximumFractionDigits(2);
    textField = new TextField();
    slider = new Slider(0, MAX, 0);
    comboBox = new ComboBox(units);
    comboBox.setConverter(new StringConverter<Unit>() {
        @Override
        public String toString(Unit t) {
            return t.description;
        }
        @Override
        public Unit fromString(String string) {
            throw new UnsupportedOperationException ( " Not
supported yet.");
        }
    })
    VBox vbox = new VBox(textField, slider);
    HBox hbox = new HBox(vbox, comboBox);
    setContent(hbox);
    this.meters = meters;
    comboBox.getSelectionModel().select(0);
}
```

注意: 最后一行代码选择 ComboBox 对象中的一个值。

3）创建 InvalidationListener 对象

要监听文本字段和组合框属性的更改，可以用 Meters 和 toMeters 创建 InvalizationListener 对象，如 Example 9-15 所示。

Example 9-15

```
import javafx.beans.InvalidationListener;
private InvalidationListener fromMeters = t -> {
    if (!textField.isFocused()) {
        textField.setText(numberFormat.format(meters.get() / getMultiplier()));
    }
};
private InvalidationListener toMeters = t -> {
    if (!textField.isFocused()) {
        return;
        try {
            meters.set(numberFormat.parse(textField.getText()).doubleValue() * getMultiplier());
        } catch (ParseException | Error | RuntimeException ignored) {
        }
    };
```

4）向控件添加更改侦听器并确保同步

要在文本字段和组合框之间保持同步，可以添加更改侦听器，如 Example 9-16 所示。

Example 9-16

```
meters.addListener(fromMeters);
comboBox.valueProperty().addListener(fromMeters);
textField.textProperty().addListener(toMeters);
fromMeters.invalidated(null);
```

在滑块的值和 meters 对象的值之间创建双向绑定，如 Example 9-17 所示。

Example 9-17

```
slider.valueProperty().bindBidirectional(meters);
```

在文本字段中输入新值时，将调用 toMeters 侦听器的 invalidated() 方法，该方法将更新 meters 对象的值。

9. 创建转换器类

在 IDE 中打开 Converter.java 文件。查看由 NetBeans IDE 自动生成的 Java 文件，删除 main 方法之外的所有代码，然后按 Ctrl＋Shift＋I 键更正导入语句。

1) 定义实例变量

为 ObservableList、DoubleProperty 简化子属性类，并创建适当类型的 MetricDistance、UsaDistance 和 meters 变量，如 Example 9-18 所示。

Example 9-18

```
import javafx.beans.property.DoubleProperty;
import javafx.collections.ObservableList;
import javafx.beans.property.SimpleDoubleProperty;
private ObservableList<Unit> metricDistances;
private ObservableList<Unit> usaDistances;
private DoubleProperty meters = new SimpleDoubleProperty(1);
```

2) 为转换器类创建构造方法

在 Converter 类的构造方法中，为度量和美制的距离创建单位对象，如 Example 9-19 所示。为 FXCollections 类添加导入语句，然后将用这些单位实例化两个 ConversionPanel 对象。

Example 9-19

```
import javafx.collections.FXCollections;
public Converter() {
metricDistances = FXCollections.observableArrayList(
new Unit("Centimeters", 0.01),
new Unit("Meters", 1.0),
new Unit("Kilometers", 1000.0));
usaDistances = FXCollections.observableArrayList(
new Unit("Inches", 0.0254),
new Unit("Feet", 0.305),
new Unit("Yards", 0.914),
new Unit("Miles", 1613.0));
}
```

3) 创建图形场景

重写 start()方法，为转换器创建图形场景应用，将两个 ConversionPanel 对象添加到图形场景中,并垂直放置它们。注意:两个 ConversionPanel 对象是用同一个 meters 对象实例化的。将 VBox 类作为图形场景的根容器。实例化两个 ConversionPanel 对象,如 Example 9-20 所示。

Example 9-20

```
@Override
public void start(Stage stage) {
    VBox vbox = new VBox(new ConversionPanel("Metric System",
metricDistances, meters), new ConversionPanel("U.S. System",
usaDistances, meters));
    Scene scene = new Scene(vbox);
    stage.setTitle("Converter");
    stage.setScene(scene);
    stage.show();
}
```

JavaFX 中的转换器应用程序的执行结果如图 9.4 所示。

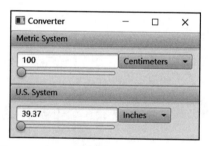

图 9.4　JavaFX 转换器应用程序的执行结果

比较使用 Swing 库和 JavaFX 实现相同功能的两个应用程序。与 Swing 应用程序的 5 个文件相比,JavaFX 中的应用程序仅包含 3 个文件,而且 JavaFX 中的代码更干净,应用程序在外观和感觉上也有所不同。

Application Files(Source Code)
- Converter.java
- ConversionPanel.java
- NetBeans Projects

9.5 小结

本章概述了JavaFX对GUI开发人员的技术优势，介绍了基于Swing组件的JavaFX应用开发，探讨了如何将JavaFX内容集成到Swing应用中，以及如何在JavaFX应用中使用Swing组件，通过两个综合性的示例介绍了如何基于Swing组件进行JavaFX应用的开发。

Chapter 10
第 10 章　基于 JavaFX 的图表应用开发

图表是许多业务应用的重要方面。在 JavaFX 中，包含一个用于创建图表的 API。因为图表是一个节点，所以可以将图表与 JavaFX 应用的其他部分集成在一起。因此，图表是 JavaFX 业务应用不可或缺的一部分。Chart API 包含许多方法，允许开发人员更改图表的外观、视觉以及数据，使其成为一个易于扩展且灵活的 API，而且这些设置的默认值非常合理，只需要几行代码就可以将图表与自定义的应用集成。JavaFX 9 中的图表 API 有 8 个具体的实现，可供开发人员使用。除此之外，开发人员还可以通过扩展一个抽象类添加自己的实现。

10.1　JavaFX 图表 API 的结构

现实世界中存在着不同类型的图表，并且有多种方法对其进行分类。JavaFX 图表 API 分为双轴图表和无轴图表。JavaFX 9 包含一个无轴图表的实现，即 PieChart。许多的双轴图表都扩展了抽象的 XYChart 类，如图 10.1 所示。

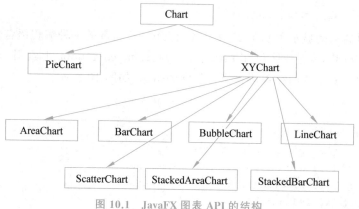

图 10.1　JavaFX 图表 API 的结构

图表由 3 部分组成：标题、图例和内容。内容针对图表的每个实现都是特定的，但图例和标题的概念在各个实现中是相似的。因此，Chart 类有许多属性以及相应的 getter( ) 和 setter( ) 方法允许对这些概念进行操作。Chart 类在 JavaDoc 中定义了如下属性。

```
BooleanProperty animated;
ObjectProperty<Node> legend;
BooleanProperty legendVisible;
ObjectProperty<Side> legendSide;
StringProperty title;
ObjectProperty<Side> titleSide;
```

接下来的示例中使用了其中的一些属性，也展示了即使没有设置这些属性的值，图表 API 也允许创建图表。因为图表扩展了区域、父节点和节点，所以这些表上可用的所有属性和方法也可以用在图表上。其中一个好处是相同的 CSS 样式技术用于向 JavaFX 节点添加样式信息，同样也适用于 JavaFX 图表。JavaFX CSS 参考指南（http://download.java.net/jdk8/jfx docs/javafx/Scene/doc/File/cssref.html）中包含可由设计师和开发者更改的 CSS 属性概述。在默认情况下，JavaFX 9 运行时附带的样式表同样适用于 JavaFX 图表。有关在 JavaFX 图表中使用 CSS 样式的更多信息，请参阅 Oracle 官网的图表教程。

10.2 使用 JavaFX PieChart

饼图以典型的饼状结构呈现信息，其中切片的大小与数据的值呈比例。在深入探讨细节之前，首先展示一个呈现 PieChart 的应用。

1. PieChart 示例

下面的示例显示了基于 2017 年 4 月 TIOBE 指数的多种编程语言的市场占有率。TIOBE 编程社区可在 https://www.tiobe.com/tiobe 上索引，它提供了基于搜索引擎流量的编程语言流行程度的指示，如图 10.2 所示。

Example 10-1　Rendering the TIOBE Index in a PieChart

```
package com.projavafx.charts;
import javafx.application.Application;
import javafx.collections.FXCollections;
import javafx.collections.ObservableList;
import javafx.scene.Scene;
```

第 10 章 基于 JavaFX 的图表应用开发

Apr 2017	Apr 2016	Change	Programming Language	Ratings	Change
1	1		Java	15.568%	-5.28%
2	2		C	6.966%	-6.94%
3	3		C++	4.554%	-1.36%
4	4		C#	3.579%	-0.22%
5	5		Python	3.457%	+0.13%
6	6		PHP	3.376%	+0.38%
7	10	∧	Visual Basic .NET	3.251%	+0.98%
8	7	∨	JavaScript	2.851%	+0.28%
9	11	∧	Delphi/Object Pascal	2.816%	+0.60%
10	8	∨	Perl	2.413%	-0.11%
11	9	∨	Ruby	2.310%	-0.04%
12	15	∧	Swift	2.287%	+0.81%
13	12	∨	Assembly language	2.168%	-0.03%
14	13	∨	Objective-C	2.163%	+0.45%
15	18	∧	R	2.138%	+0.87%
16	14	∨	Visual Basic	2.058%	+0.45%
17	16	∨	MATLAB	2.045%	+0.70%
18	44	⇑	Go	1.974%	+1.73%
19	24	⇑	Scratch	1.668%	+0.86%
20	17	∨	PL/SQL	1.619%	+0.30%

图 10.2 2017 年 4 月 TIOBE 编程社区索引

```
import javafx.scene.chart.PieChart;
import javafx.scene.layout.StackPane;
import javafx.stage.Stage;
public class ChartApp1 extends Application {
    @Override
    public void start(Stage primaryStage) {
        PieChart pieChart = new PieChart();
        pieChart.setData(getChartData());
        primaryStage.setTitle("PieChart");
        StackPane root = new StackPane();
        root.getChildren().add(pieChart);
        primaryStage.setScene(new Scene(root, 400, 250));
        primaryStage.show();
    }
    private ObservableList<PieChart.Data> getChartData() {
        ObservableList < PieChart. Data > answer = FXCollections.observableArrayList();
```

```
            answer.addAll(new PieChart.Data("java", 15.57), new
PieChart.Data("C", 6.97), new PieChart.Data("C++", 4.55), new
PieChart.Data("C#", 3.58), new PieChart.Data("Python", 3.45), new
PieChart.Data("PHP", 3.38), new PieChart.Data("Visual Basic .NET", 3.
25));
        return answer;
    }
    public static void main(String[] args) {
        launch(args);
    }
}
```

该示例的运行结果如图 10.3 所示。

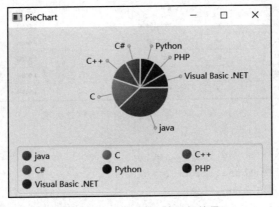

图 10.3　ChartApp1 的运行结果

【分析讨论】

- 首先,设置应用程序、舞台和场景所需的代码;然后,PieChart 扩展了节点,因此可以将其添加到场景图中。
- start()方法中的前两行代码创建了 PieChart,并向其中添加了所需的数据:

```
PieChart pieChart = new PieChart();
pieChart.setData(getChartData());
```

- 数据类型为 ObservableList<PieChart.Data>,是用 getChartData()方法获得的,在示例中,它包含静态数据。例如,getChartData()方法的返回类型所指的就是静态数据,返回的数据是 PieChart 的 ObservableList。

第 10 章 基于 JavaFX 的图表应用开发

Data,是 PieChart 的一个嵌套类,它包含绘制饼图一部分所需的信息。
- Data 有一个构造方法,该构造方法采用切片的名称及其值。使用这个构造方法可以创建包含编程语言名称及其在 TIOBE 索引中的分数的数据元素。

```
new PieChart.Data("java", 15.57)
```

然后,将这些数据元素添加到 ObservableList＜PieChart.Data＞中并返回。

2. 完善 PieChart 示例

虽然这个示例的结果看起来不错,但仍然可以进行代码调整和渲染。
该示例使用两行代码创建 PieChart 并用数据填充它。

```
PieChart pieChart = new PieChart();
pieChart.setData(getChartData());
```

因为 PieChart 也有一个单参数的构造方法,所以上述代码片段可以按照以下方式进行替换。

```
PieChart pieChart = new PieChart(getChartData());
```

除了抽象图表类定义的属性外,PieChart 还具有以下属性。

```
BooleanProperty clockwise
ObjectProperty<ObservableList<PieChart.Data>> data
DoubleProperty labelLineLength
BooleanProperty labelsVisible
DoubleProperty startAngle
```

10.1 节介绍了数据属性。其他属性将在下一个代码片段中展示。Example 10-2 包含 start( )方法的修改版本。

Example 10-2　Modified Version of the PieChart Example

```
public void start(Stage primaryStage) {
    PieChart pieChart = new PieChart();
    pieChart.setData(getChartData());
    pieChart.setTitle("Tiobe index");
    pieChart.setLegendSide(Side.LEFT);
```

```
        pieChart.setClockwise(false);
        pieChart.setLabelsVisible(false);
        primaryStage.setTitle("PieChart");
        StackPane root = new StackPane();
        root.getChildren().add(pieChart);
        primaryStage.setScene(new Scene(root, 400, 250));
        primaryStage.show();
    }
```

因为这里使用了侧边,所以新代码的左字段中必须在应用中导入 Side 类,这是通过在代码的导入块中添加以下代码完成的。

```
import javafx.geometry.Side;
```

运行修改后的代码会得到新的输出结果,如图 10.4 所示。

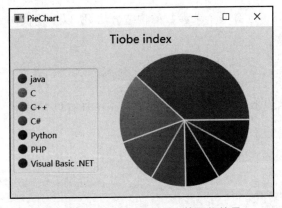

图 10.4 修改后 ChartApp1 的运行结果

更改几行代码会导致输出结果看起来非常不同。下面详细阐述所做的更改。

(1) 在图表中添加了一个标题,这是通过下列语句实现的。

```
pieChart.setTitle("Tiobe index");
```

或者使用下列语句:

```
pieChart.titleProperty().set("Tiobe index");
```

这两种方法的结果相同。注意:接下来的修改也可以使用相同的模式完成,这里只使用了 setter()方法,但很容易用基于属性的方法替换。

(2)修改后的示例中的下一行代码更改了图例的位置:

```
pieChart.setLegendSide(Side.LEFT);
```

如果未指定 legendSide,则图例将显示在图表下方的默认位置。标题和 legendSide 都是属于抽象图表类的属性,因此它们可以设置在任何图表上。

(3)修改后的示例中的下一行修改了特定于 pieChart 的属性:

```
pieChart.setClockwise(false);
```

在默认情况下,PieChart 中的切片是顺时针渲染的。通过将此属性设置为 false 可以使切片逆时针渲染,并且禁用了在图表中显示标签。标签仍显示在图例中,但它们不再指向各个切片,这是通过以下代码实现的:

```
pieChart.setLabelsVisible(false);
```

(4)应用程序,尤其是使用 CSS 样式表的图表。从 Java 代码中删除布局更改,并添加一个包含一些布局说明的样式表。Example 10-3 显示了 start()方法的修改代码,Example 10-4 包含添加的样式表。

Example 10-3　Remove Programmatic Layout Instructions

```
public void start(Stage primaryStage) {
    PieChart pieChart = new PieChart();
    pieChart.setData(getChartData());
    pieChart.titleProperty().set("Tiobe index");
    primaryStage.setTitle("PieChart");
    StackPane root = new StackPane();
    root.getChildren().add(pieChart);
    Scene scene = new Scene (root, 400, 250);
    scene.getStylesheets().add("/chartappstyle.css");
    primaryStage.setScene(scene);
    primaryStage.show();
}
```

Example 10-4　Style Sheet for PieChart Example

```
.chart {
```

```
-fx-clockwise: false;
-fx-pie-label-visible: true;
-fx-label-line-length: 5;
-fx-start-angle: 90;
-fx-legend-side: right;
}
.chart-pie-label {
-fx-font-size:9px;
}
.chart-content {
-fx-padding:1;
}
.default-color0.chart-pie {
-fx-pie-color:blue;
}
```

经过上述修改后的 ChartApp1 的运行结果如图 10.5 所示。

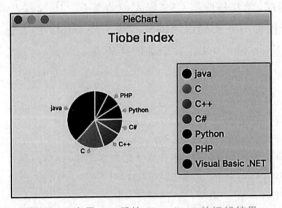

图 10.5　应用 CSS 后的 ChartApp1 的运行结果

现在回顾所做的更改。在详细讨论各个更改之前，首先展示如何将 CSS 包含在应用程序中，这是通过将样式表添加到场景中实现的，实现语句如下：

```
scene.getStylesheets().add("/chartappstyle.css");
```

在运行时，样式表文件必须包含在 classpsath 环境变量设置的值中，并且设置 Clockwise 为 false。

```
pieChart.setClockwise(false);
```

从 Example 10-3 的代码中删除了这一行,而且在样式表中的 Chart 类上定义了-fx 顺时针属性:

```
.chart {
    -fx-clockwise: false;
    -fx-pie-label-visible: true;
    -fx-label-line-length: 5;
    -fx-start-angle: 90;
    -fx-legend-side: right;
}
```

同样,在图表类的定义中,通过将-fx-pie-label-visible 属性设置为 true,使饼图上的标签可见,并将每个标签的线条长度指定为 5。此外,将整个饼图旋转 90°,这是通过定义-fx start- angle 属性实现的。标签现在在样式表中定义,通过省略以下行从代码中删除相应的定义。

```
pieChart.setLabelsVisible(false)
```

为了确保图例显示在图表的右侧,指定了-fx-legend-side 属性。默认情况下,PieChart 使用 caspian 样式表中定义的默认颜色。第一个切片用 default-color0 填充,第二个切片用 default-color1 填充,以此类推。更改不同切片颜色最简单的方法是覆盖默认颜色的定义。在样式表中,这是由以下语句实现的:

```
.default-color0.chart-pie {
    -fx-pie-color: blue;
}
```

其他切片也可以这样做。如果在没有 CSS 其他部分的情况下运行示例,会注意到图表本身非常小,标签占用了太多的空间。因此,修改标签的字体大小如下:

```
.chart-pie-label {
    -fx-font-size:9px;
}
```

此外,还减少了图表区域中的填充:

```
.chart-content {
    -fx-padding:1;
}
```

最后,改变了背景、边框颜色以及宽度,这是通过覆盖图表图例类实现的,如下所示。

```
.chart-legend {
  -fx-background-color:#f0e68c;
  -fx-border-color:#696969;
  -fx-border-width:1;
}
```

10.3 使用 XYChart

XYChart 类是一个抽象类,它有 7 个直接已知的子类。这些类与 PieChart 类的区别在于,XYChart 有两个轴和可选的 alternativeColumn 或 alternativeRow,这将转换为 XYChart 上的以下附加属性列表。

```
BooleanProperty alternativeColumnFillVisible
BooleanProperty alternativeRowFillVisible
ObjectProperty<ObservableList<XYChart.Series<X,Y>>> data
BooleanProperty horizontalGridLinesVisible
BooleanProperty horizontalZeroLineVisible
BooleanProperty verticalGridLinesVisible
BooleanProperty verticalZeroLineVisible
```

XYChart 类中的数据按顺序排列。这些系列的渲染方式是特定于 XYChart 子类的实现。通常,一个系列中的单个元素包含多个对。以下示例使用 3 种编程语言假设的市场份额预测从 2017 年 Java、C 和 C++ 的 TIOBE 索引开始,并添加随机值(介于到 2020 年,每年将它们减少-2 和增加 2)。Java 的结果对(年份、数字)构成了 Java 系列,C 和 C++ 也是如此。因此有 3 个系列,每个系列包含 10 对。

PieChart 和 XYChart 类的主要区别在于 XYChart 类中存在 x 轴和 y 轴,创建 XYChart 类时需要这些轴,可以从如下构造方法中观察到。

```
XYChart (Axis<X> xAxis, Axis<Y> yAxis)
```

Axis 类是一个抽象类扩展区域(也扩展父类和节点),包含两个子类:CategoryAxis 和 ValueAxis。CategoryAxis 用于呈现字符串格式的标签,这从类定义中可以看出:

```
public class CategoryAxis extends Axis<java.lang.String>
```

ValueAxis 类用于呈现表示数字的数据条目,它本身就是一个抽象类,定义如下。

```
public abstract class ValueAxis <T extends java.lang.Number> extends
Axis<T>
```

ValueAxis 类有一个具体的子类,即 NumberAxis:

```
public final class NumberAxis extends ValueAxis<java.lang.Number>
```

这些 Axis 类之间的差异将在示例中变得清晰。现在展示一些不同 XYChart 类实现的示例,从散点图开始。散点图部分还介绍了所有 XYChart 类的一些常见功能。

注意:因为 Axis 类扩展了区域,所以它们允许应用与其他区域相同的 CSS 元素,还允许高度订制的轴实例。

Example 10-5 展示了应用程序的第一个使用散点图的实现。

Example 10-5　First Implementation of Rendering Data in a ScatterChart

```
package com.projavafx;
import javafx.application.Application;
import javafx.collections.FXCollections;
import javafx.collections.ObservableList;
import javafx.scene.Scene;
import javafx.scene.chart.NumberAxis;
import javafx.scene.chart.ScatterChart;
import javafx.scene.chart.XYChart;
import javafx.scene.chart.XYChart.Series;
import javafx.scene.layout.StackPane;
import javafx.stage.Stage;
public class ChartApp3 extends Application {
    public static void main(String[] args) {
        launch(args);
    }
    @Override
    public void start(Stage primaryStage) {
```

```
            NumberAxis xAxis = new NumberAxis();
            NumberAxis yAxis = new NumberAxis();
            ScatterChart scatterChart = new ScatterChart(xAxis, yAxis);
            scatterChart.setData(getChartData());
            primaryStage.setTitle("ScatterChart");
            StackPane root = new StackPane();
            root.getChildren().add(scatterChart);
            primaryStage.setScene(new Scene(root, 400, 250));
            primaryStage.show();
        }
        private ObservableList< XYChart.Series< Integer, Double > > getChartData() {
            double javaValue = 15.57;
            double cValue = 6.97;
            double cppValue = 4.55;
            ObservableList< XYChart.Series< Integer, Double > > answer = FXCollections.observableArrayList();
            Series<Integer, Double> java = new Series<>();
            Series<Integer, Double> c = new Series<>();
            Series<Integer, Double> cpp = new Series<>();
            for (int i = 2017; i < 2027; i++) {
                java.getData().add(new XYChart.Data(i, javaValue));
                javaValue = javaValue + 4 * Math.random() - 2;
                c.getData().add(new XYChart.Data(i, cValue));
                cValue = cValue + Math.random() - .5;
                cpp.getData().add(new XYChart.Data(i, cppValue));
                cppValue = cppValue + 4 * Math.random() - 2;
            }
            answer.addAll(java, c, cpp);
            return answer;
        }
    }
```

运行结果如图 10.6 所示。

虽然图表显示了所需的信息，但是可读性不强，还需要添加一些增强功能，与 PieChart 类示例类似，它创建了一个单独的方法获取数据，其原因是在现实世界的应用程序中不太可能有静态数据。通过隔离数据并用单独的方法检索，使得更改获取数据的方式更加容易。单个数据点由 XYChart 实例定义。其中，

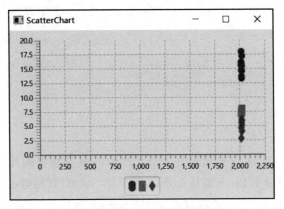

图 10.6 ChartApp3 的运行结果

参数具有以下定义。

```
i: Integer, representing a specific year (between 2017 and 2026)
d: Double, representing the hypothetical TIOBE index for the
particular series in the year specified by I
```

单个数据点由 XYChart 实例定义——数据＜Integer,Double＞表示使用的是一张图表。XYChart.Series＜Integer,Double＞中的参数具有以下定义。

```
java.getData().add(...)
c.getData().add(...)
and
cpp.getData().add(...)
```

最后,所有系列都被添加到 ObservableList＜XYChart.Series＜Integer,Double＞＞中并返回。应用程序的 start() 方法包含创建和呈现散点图以及使用从 getChartData() 方法中获得的数据填充散点图所需的功能。

如前所述,可以在 PieChart 类中使用不同的模式,示例中使用了 JavaBeans 模式,但也可以使用属性。

要创建散点图,需要创建一个 xAxis 和一个 yAxis。在第一个示例中,使用了 NumberAxis 的两个实例:

```
NumberAxis xAxis = new NumberAxis();
NumberAxis yAxis = new NumberAxis();
```

除了调用下面的 ScatterChart() 构造方法之外，这个方法与 PieChart 类并没有什么不同。

```
ScatterChart scatterChart = new ScatterChart(xAxis, yAxis);
```

10.4　改进示例的实现

观察图 10.6，首先观察到的是一个系列中的所有数据图几乎都是渲染在彼此之上的。原因很明显：x 轴从 0 开始，到 2250 结束，默认情况下，NumberAxis 会自动确定其范围。可以通过设置 autoRanging 属性为 false，并提供下限和上限的值实现。如果在原始示例中进行更换，则可以通过以下代码片段实现：

```
NumberAxis xAxis = new NumberAxis();
xAxis.setAutoRanging(false);
xAxis.setLowerBound(2017);
xAxis.setUpperBound(2027);
```

运行结果如图 10.7 所示。

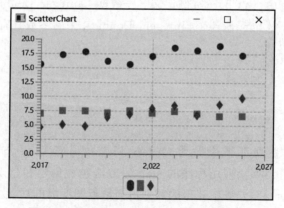

图 10.7　改进后的 ChartApp3 的运行结果（1）

可以向 XYChart 类的 3 个实例添加名称，也可以向图例节点中的符号添加标签。getChartData() 方法的相关部分变为：

```
Series<Integer, Double> java = new Series<>();
Series<Integer, Double> c = new Series<>();
```

```
Series<Integer, Double> cpp = new Series<>();
java.setName("java");
c.setName("C");
cpp.setName("C++");
```

运行结果如图 10.8 所示。

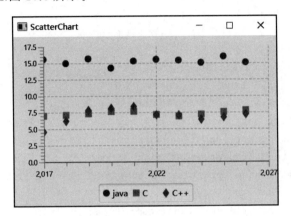

图 10.8　改进后的 ChartApp3 的运行结果（2）

目前用数字 X 表示 X。因为年份可以用数字表示，所以这是有效的。由于不对年份进行任何数字运算，而且连续输入数据之间的距离始终为一年，因此也可以使用字符串值表示此信息。现在修改代码，以使用 xAxis 的 CategoryAxis。将 xAxis 从 NumberAxis 更改为 CategoryAxis 还意味着 getChartData( )方法应该返回 ObservableList＜XYChart.Series＜String，Double＞＞，这意味着单个系列中的不同元素应该具有 XYChart 类型——数据＜String，Double＞。在 Example 10-6 中，原始代码已被修改为使用 CategoryAxis。

Example 10-6　Using CategoryAxis Instead of NumberAxis for the xAxis

```
package projavafx;
import javafx.application.Application;
import javafx.collections.FXCollections;
import javafx.collections.ObservableList;
import javafx.scene.Scene;
import javafx.scene.chart.CategoryAxis;
import javafx.scene.chart.NumberAxis;
import javafx.scene.chart.ScatterChart;
```

```java
import javafx.scene.chart.XYChart;
import javafx.scene.chart.XYChart.Series;
import javafx.scene.layout.StackPane;
import javafx.stage.Stage;
public class ChartApp4 extends Application {
    public static void main(String[ ] args) {
        launch(args);
    }
    @Override
    public void start(Stage primaryStage) {
        CategoryAxis xAxis = new CategoryAxis( );
        NumberAxis yAxis = new NumberAxis( );
        ScatterChart scatterChart = new ScatterChart(xAxis, yAxis);
        scatterChart.setData(getChartData( ));
        scatterChart.setTitle("speculations");
        primaryStage.setTitle("ScatterChart example");
        StackPane root = new StackPane( );
        root.getChildren( ).add(scatterChart);
        primaryStage.setScene(new Scene(root, 400, 250));
        primaryStage.show( );
    }
     private  ObservableList < XYChart. Series < String,  Double > > getChartData( ) {
        double javaValue = 15.57;
        double cValue = 6.97;
        double cppValue = 4.55;
        ObservableList< XYChart. Series< String, Double> > answer = FXCollections.observableArrayList( );
        Series<String, Double> java = new Series<>( );
        Series<String, Double> c = new Series<>( );
        Series<String, Double> cpp = new Series<>( );
        java.setName("java");
        c.setName("C");
        cpp.setName("C++");
        for (int i = 2017; i < 2027; i++) {
           java.getData().add(new XYChart.Data(Integer.toString(i), javaValue));
            javaValue = javaValue + 4 * Math.random() - .2;
```

```
            c.getData().add(new XYChart.Data(Integer.toString(i),
cValue));
            cValue = cValue + 4 * Math.random() - 2;
            cpp.getData().add(new XYChart.Data(Integer.toString(i),
cppValue));
            cppValue = cppValue + 4 * Math.random() - 2;
        }
        answer.addAll(java, c, cpp);
        return answer;
    }
}
```

运行结果如图 10.9 所示。

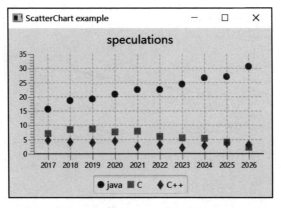

图 10.9　ChartApp4 的运行结果

10.5　使用 LineChart

10.4 节中的示例将数据条目用单点或符号表示。通常情况下，人们希望用一条线将点连接起来，因为这样有助于看到趋势。JavaFX 折线图非常适合这种情况。线形图的 API 与散点图的 API 有许多共同的方法。事实上，可以重用 Example 10-6 中的大部分代码，只需要将散点图替换为折线图即可。因为数据仍然完全相同，所以只在 Example 10-7 中显示新的 start( )方法。

Example 10-7　Using a LineChart Instead of a ScatterChart

```
public void start(Stage primaryStage) {
```

```
            CategoryAxis xAxis = new CategoryAxis();
            NumberAxis yAxis = new NumberAxis();
            LineChart lineChart = new LineChart(xAxis, yAxis);
            lineChart.setData(getChartData());
            lineChart.setTitle("speculations");
            primaryStage.setTitle("LineChart example");
            StackPane root = new StackPane();
            root.getChildren().add(lineChart);
            primaryStage.setScene(new Scene(root, 400, 250));
            primaryStage.show();
        }
```

运行结果如图 10.10 所示。

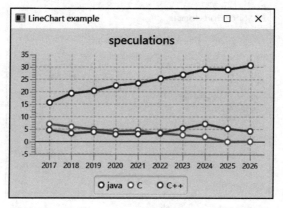

图 10.10　使用 LineChart 的 ChartApp4 的运行结果

散点图的大部分可用功能也可用于折线图。使用折线图可以更改图例的位置、添加或删除标题，以及使用数字轴而非类别轴。

10.6　使用 BarChart

条形图能够呈现与散点图和折线图相同的数据。在条形图中，重点通常更多的是显示给定类别的不同系列之间的相对差异。在下面的示例中，这意味着需要关注 Java、C 以及 C++。同样，可以不需要修改返回数据的方法。事实上，条形图需要 CategoryAxis 为其 xAxis，这是因为已经修改了 getChartData() 方法以返回包含 XYChart 的可观察列表。从 Example 10-6 开始，只从散点图更改到条形图，于是可以得到 Example 10-8。

Example 10-8　Using a BarChart Instead of a ScatterChart

```
public void start(Stage primaryStage) {
    CategoryAxis xAxis = new CategoryAxis();
    NumberAxis yAxis = new NumberAxis();
    BarChart barChart = new BarChart(xAxis, yAxis);
    barChart.setData(getChartData());
    barChart.setTitle("speculations");
    primaryStage.setTitle("BarChart example");
    StackPane root = new StackPane();
    root.getChildren().add(barChart);
    primaryStage.setScene(new Scene(root, 400, 250));
    primaryStage.show();
}
```

下面把 JavaFX 的"import javafx.scene.chart.ScatterChart;"语句替换成"import javafx.scene.chart. BarChart;",再次构建应用程序并运行它,结果是一个如图 10.11 所示的条形图。

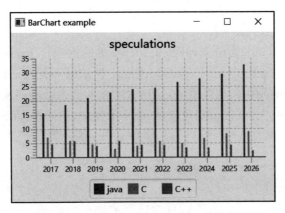

图 10.11　使用 BarChart 的 ChartApp4 的运行结果(1)

虽然结果确实显示了每年的数值之间的差异,但这并不是十分清楚,因为界限相当小。由于总场景宽度为 400 像素,因此没有太多空间渲染大型条形图。但是,条形图 API 包含定义条形图之间内部间隙和类别之间间隙的方法。在这个示例中,我们希望条形之间的间隙更小,例如 1 像素,这可以通过调用完成:

```
barChart.setBarGap(1);
```

将这一行代码添加到 start()方法中,然后重新运行程序,运行结果如图 10.12 所示。

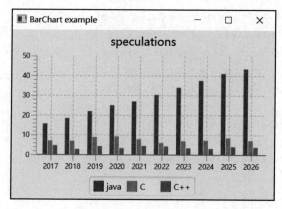

图 10.12　使用 BarChart 的 ChartApp4 的运行结果(2)

显然,这一行代码会导致可读性上的巨大差异。

10.7　使用 StackedBarChart

StackedBarChart 是在 JavaFX 2.1 中新增的内容。StackedBarChart 以条形图的形式显示数据,但是 StackedBarChart 显示的不是相邻的同一类别的条形图,而是同一类别内的相互重叠,这通常会使检查总数变得更容易。通常,类别与数据系列中的公共键值相对应,因此在示例中,不同年份(2017 年、2018 年、2026 年)可以被视为类别。可以加上这些 xAxis 的分类如下:

```
IntStream.range(2017,2026).forEach(t -> xAxis.getCategories().add
(String.valueOf(t)));
```

除此之外,唯一的代码更改是在 import 语句中用 StackedBarChart 替换条形图。如 Example 10-9 中的代码片段。

Example 10-9　Using a StackedBarChart Instead of a ScatterChart

```
public void start(Stage primaryStage) {
    CategoryAxis xAxis = new CategoryAxis();
    IntStream.range(2017,2026).forEach(t -> xAxis.getCategories().
add(String.valueOf(t)));
    NumberAxis yAxis = new NumberAxis();
```

```
        StackedBarChart stackedBarChart = new StackedBarChart(xAxis,
yAxis, getChartData());
    stackedBarChart.setTitle("speculations");
    primaryStage.setTitle("StackedBarChart example");
    StackPane root = new StackPane();
    root.getChildren().add(stackedBarChart);
    Scene scene = new Scene(root, 400, 250);
    primaryStage.setScene(scene);
    primaryStage.show();
}
```

运行结果如图 10.13 所示。

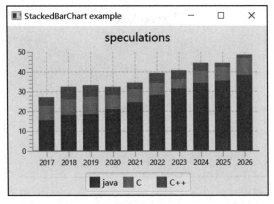

图 10.13　使用 StackedBarChart 的 ChartApp4 的运行结果

10.8　使用 AreaChart

在某些情况下,填充连接点的线下区域是有意义的。虽然呈现的数据与折线图相同,但结果看起来却不同。Example 10-10 中包含修改后的 start( )方法,该方法使用面积图而非原始散点图。与前面的修改一样,并没有更改 getChartData( )方法。

Example 10-10　Using an AreaChart Instead of a ScatterChart

```
public void start(Stage primaryStage) {
    CategoryAxis xAxis = new CategoryAxis();
    NumberAxis yAxis = new NumberAxis();
```

```
        AreaChart areaChart = new AreaChart(xAxis, yAxis);
        areaChart.setData(getChartData());
        areaChart.setTitle("speculations");
        primaryStage.setTitle("AreaChart example");
        StackPane root = new StackPane();
        root.getChildren().add(areaChart);
        primaryStage.setScene(new Scene(root, 400, 250));
        primaryStage.show();
    }
```

运行结果如图 10.14 所示。

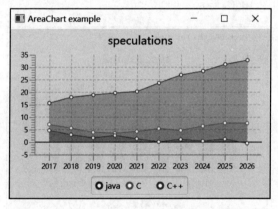

图 10.14 使用 AreaChart 的 ChartApp4 的运行结果

10.9 使用 StackedAreaChart

堆叠条形图与面积图的关系就像堆叠条形图与条形图的关系一样。StackedAreaChart 始终显示特定类别中的值之和,而不是显示单个区域。将面积图更改为 StackedAreaChart,只需要更改一行代码和相应的导入语句,即将下列这条语句:

```
        AreaChart areaChart = new AreaChart(xAxis, yAxis);
```

替换成

```
        StackedAreaChart areaChart = new StackedAreaChart(xAxis, yAxis);
```

运行结果如图 10.15 所示。

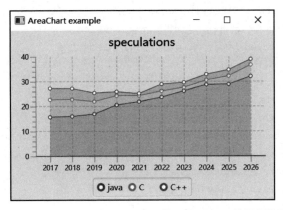

图 10.15　使用 StackedAreaChart 的 ChartApp4 的运行结果

10.10　使用 BubbleChart

XYChart 类的最后一个实现是一个特殊的实现。BubbleChart 不包含在 XYChart 类中，但它是当前 JavaFX 图表 API 中唯一能直接实现使用 XYChart 类上的附加参数的数据类。

修改 Example 10-6 中的代码，使用 BubbleChart 代替散点图。默认情况下，当 X 轴上的跨度与 Y 轴上的跨度大不相同时，气泡会被拉伸，这里不使用"年"，而是使用"十分之一年"作为 xAxis 上的值。这样，就有了 100 个单元跨度的 xAxis(10 年)与 yAxis 上约 30 个单元跨度的比较，这或多或少也是图表的宽度和高度之间的一个比例。因此，气泡是相对圆形的。Example 10-11 包含呈现 BubbleChart 的代码。

Example 10-11　Using the BubbleChart

```
package com.projavafx.charts;
import javafx.application.Application;
import javafx.collections.FXCollections;
import javafx.collections.ObservableList;
import javafx.scene.Scene;
import javafx.scene.chart.*;
import javafx.scene.chart.XYChart.Series;
import javafx.scene.layout.StackPane;
```

```java
import javafx.stage.Stage;
import javafx.util.StringConverter;
public class ChartApp5 extends Application {
    public static void main(String[ ] args) {
        launch(args);
    }
    @Override
    public void start(Stage primaryStage) {
        NumberAxis xAxis = new NumberAxis( );
        NumberAxis yAxis = new NumberAxis( );
        yAxis.setAutoRanging(false);
        yAxis.setLowerBound(0);
        yAxis.setUpperBound(30);
        xAxis.setAutoRanging(false);
        xAxis.setAutoRanging(false);
        xAxis.setLowerBound(20170);
        xAxis.setUpperBound(20261);
        xAxis.setTickUnit(10);
        xAxis.setTickLabelFormatter(new StringConverter<Number>( ) {
            @Override
            public String toString(Number n) {
                return String.valueOf(n.intValue( ) / 10);
            }
            @Override
            public Number fromString(String s) {
                return Integer.valueOf(s) * 10;
            }
        });
        BubbleChart bubbleChart = new BubbleChart(xAxis, yAxis);
        bubbleChart.setData(getChartData( ));
        bubbleChart.setTitle("Speculations");
        primaryStage.setTitle("BubbleChart example");
        StackPane root = new StackPane( );
        root.getChildren( ).add(bubbleChart);
        primaryStage.setScene(new Scene(root, 400, 250));
        primaryStage.show( );
    }
    private ObservableList < XYChart.Series < Integer, Double > > getChartData( ) {
```

```
            double javaValue = 15.57;
            double cValue = 6.97;
            double cppValue = 4.55;
            ObservableList<XYChart.Series<Integer, Double>> answer =
    FXCollections.observableArrayList();
            Series<Integer, Double> java = new Series<>();
            Series<Integer, Double> c = new Series<>();
            Series<Integer, Double> cpp = new Series<>();
            java.setName("java");
            c.setName("C");
            cpp.setName("C++");
            for (int i = 20170; i < 20260; i = i + 10) {
                double diff = Math.random();
                java.getData().add(new XYChart.Data(i, javaValue));
                javaValue = Math.max(javaValue + 2 * diff - 1, 0);
                diff = Math.random();
                c.getData().add(new XYChart.Data(i, cValue));
                cValue = Math.max(cValue + 2 * diff - 1, 0);
                diff = Math.random();
                cpp.getData().add(new XYChart.Data(i, cppValue));
                cppValue = Math.max(cppValue + 2 * diff - 1, 0);
            }
            answer.addAll(java, c, cpp);
            return answer;
        }
    }
```

xAxis 的范围为从 2017 年到 2026 年，但是如果想展示轴心的年份，则可以通过以下调用实现。

```
xAxis.setTickLabelFormatter(new StringConverter<Number>() {
    ...
}
```

这里提供的 StringConverter 将使用的数字（如 20210）转换为字符串（如 2021），反之亦然。这样一来，就可以使用想要的任何数量计算气泡了，并且可以很好地格式化标签。运行该示例将得到如图 10.16 所示的图表。

现在，还没有利用 XYChart 类的三参数构造方法。除了已经熟悉的两个参数构造方法：

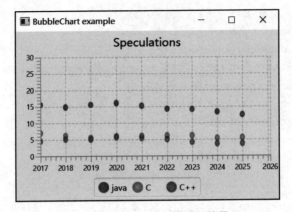

图 10.16 ChartApp5 的运行结果

```
XYChart.Data (X xValue, Y yValue)
```

XYChart 类还有一个单个参数的构造方法：

```
XYChart.Data (X xValue, Y yValue, Object extraValue)
```

extraValue 参数可以是任意类型，以允许开发人员实现自订功能的 XYChart 的子类，它利用可以包含在单个数据中的附加信息要素 BubbleChart 使用这个额外的值，以决定气泡应该有多大被渲染。

现在修改 getChartData( )方法以使用三参数构造函数。X 值和 yValue 参数仍然与上一个示例中的相同，但现在添加了第 3 个参数，它用来预示一个即将到来的趋势。这个参数越大，下一年的上升幅度就越大。修改后的 getChartData( )方法如 Example 10-12 所示。

Example 10-12 Using a Three-Argument Constructor for XYChart.Data Instances

```
private ObservableList < XYChart. Series < Integer, Double > >
getChartData( ) {
    double javaValue = 15.57;
    double cValue = 6.97;
    double cppValue = 4.55;
    ObservableList < XYChart. Series < Integer, Double > > answer =
FXCollections.observableArrayList();
    Series<Integer, Double> java = new Series<>( );
```

```
        Series<Integer, Double> c = new Series<>();
        Series<Integer, Double> cpp = new Series<>();
        java.setName("java");
        c.setName("C");
        cpp.setName("C++");
        for (int i = 20170; i < 20270; i = i+10) {
            double diff = Math.random();
             java.getData().add(new XYChart.Data(i, javaValue, 2 * diff));
            javaValue = Math.max(javaValue + 2 * diff - 1,0);
            diff = Math.random();
            c.getData().add(new XYChart.Data(i, cValue, 2 * diff));
            cValue = Math.max(cValue + 2 * diff - 1,0);
            diff = Math.random();
            cpp.getData().add(new XYChart.Data(i, cppValue, 2 * diff));
            cppValue = Math.max(cppValue + 2 * diff - 1,0);
        }
        answer.addAll(java, c, cpp);
        return answer;
    }
```

将此方法与 Example 10-11 中的 start( )方法集成，会产生如图 10.17 所示的输出。

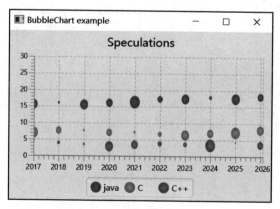

图 10.17　使用 Three-Argument Constructor for XYChart.Data 的 ChartApp5 的运行结果

关于 JavaFX Chart API 文档的详细信息，请参阅以下链接：

```
http://docs.oracle.com/javase/8/javafx/user-interface-tutorial/
charts.htm
http://docs.oracle.com/javase/8/javafx/user-interface-tutorial/
css-styles.htm
```

10.11 小结

JavaFX 图表 API 为不同的图表类型提供了许多现成的实现。每种实现都有不同的用途,可供开发人员选择最合适的图表。通过应用 CSS 规则或使用特定于图表的方法或属性,可以修改图表并针对特定的应用程序进行调整。如果需要更加自定义的图表,则可以扩展抽象图表类并利用该类上的现有属性;如果图表需要两个轴,则可以扩展抽象 XYChart 类。

第 11 章 基于 JavaFX 开发动画与视觉效果

本章将介绍如何基于 JavaFX 开发具有变换、时间轴动画以及视觉效果的 JavaFX 应用,并基于示例介绍与它们相关的概念与实现原理。

11.1 在 JavaFX 中应用变换

11.1.1 变换概述

JavaFX 支持的变换的实现位于 javafx.scene.transform 包以及 Transform 类的子类中。变换可以根据某些参数更改图形对象在坐标系中的位置。JavaFX 支持以下类型的变换:
- Translation
- Rotation
- Scaling
- Shearing

变换既可以应用于独立节点,也可以应用于多个节点组的节点。可以一次应用一个变换,也可以合并变换,并对一个节点应用多个变换。Transform 类实现了仿射变换的概念。仿射类扩展了 Transform 类,并充当所有变换的超类。仿射变换基于欧几里得代数,并通过使用矩阵执行线性映射,从初始坐标到其他坐标均保持直线度和平行度。仿射变换可以使用 ObservableArrayList 的旋转、平移、缩放和剪切实现构建。注意:通常不提倡直接使用仿射类,而是使用特定的平移、缩放、旋转或剪切的变换。

JavaFX 中的变换可以沿着 3 个坐标执行,因此用户可以创建三维(3D)对象和效果。为了管理三维图形中深度对象的显示,JavaFX 实现了 z 缓冲。z 缓冲确保了虚拟世界中的透视图与现实世界中的相同——前景中的实体阻挡了后面的实体。z 缓冲可以通过使用 setDepthTest 类实现启用。可以在示例应用程序中禁用 z 缓冲,通过执行语句"setDepthTest(DepthTest.disable)"可以查看 z 缓冲的效果。

为了简化变换使用,JavaFX 使用 x 轴、y 轴以及 x、y 和 z 轴实现变换的方法。如果要创建二维(2D)效果,则只能指定 x 和 y 坐标。如果要创建三维效果,就要指定所有的 3 个坐标。为了能够在 JavaFX 中看到三维对象和转换效果,用户必须启用透视相机。可以通过学习本书提供的示例使用转换记录,并尝试不同的转换参数。有关特定的类、方法或其他特性,请参阅 API 文档。在本书中,木琴应用程序(Xylophone.java)被用作示例,以说明所有可用的转换,运行结果如图 11.1 所示。

图 11.1　Xylophone.java 的运行结果

11.1.2　变换的类型与示例

1. 变换

平移变换相对于其初始位置的轴,将节点沿一个方向从一个位置移动到另一个位置。木琴杆的初始位置已定义了 x、y、z 的坐标。在 Example 11-1 中,初始位置值由 xStart、yPos 和 zPos 变量设置,还添加了一些其他变量以简化应用不同变换时的计算。木琴的每一小节都是基于其中一根基础杆的。然后,该示例用不同的方式转换基础钢筋并沿 3 个轴移动,以在空间中正确地定位它们。

Example 11-1　Translation

```
Group rectangleGroup = new Group();
rectangleGroup.setDepthTest(DepthTest.ENABLE);
double xStart = 260.0;
```

```
double xOffset = 30.0;
double yPos = 300.0;
double zPos = 0.0;
double barWidth = 22.0;
double barDepth = 7.0;
//Base1
Cube base1Cube = new Cube(1.0, new Color(0.2, 0.12, 0.1, 1.0), 1.0);
base1Cube.setTranslateX(xStart + 135);
base1Cube.setTranslateZ(yPos+20.0);
base1Cube.setTranslateY(11.0);
```

2. 旋转

旋转变换将围绕场景的指定轴心移动节点。可以使用 Transform 类的 rotate() 方法执行旋转。要在示例程序中围绕木琴旋转相机，可以使用旋转变换，尽管从技术上讲，旋转相机时实际上移动的是木琴本身。Example 11-2 展示了旋转变换的代码。

Example 11-2　Rotation

```
class Cam extends Group {
    Translate t = new Translate();
    Translate p = new Translate();
    Translate ip = new Translate();
    Rotate rx = new Rotate();
    { rx.setAxis(Rotate.X_AXIS); }
    Rotate ry = new Rotate();
    { ry.setAxis(Rotate.Y_AXIS); }
    Rotate rz = new Rotate();
    { rz.setAxis(Rotate.Z_AXIS); }
    Scale s = new Scale();
    public Cam() { super(); getTransforms().addAll(t, p, rx, rz, ry, s, ip); }
}
...
scene.setOnMouseDragged(new EventHandler<MouseEvent>() {
    public void handle(MouseEvent me) {
        mouseOldX = mousePosX;
        mouseOldY = mousePosY;
        mousePosX = me.getX();
```

```
            mousePosY = me.getY();
            mouseDeltaX = mousePosX - mouseOldX;
            mouseDeltaY = mousePosY - mouseOldY;
            if (me.isAltDown() && me.isShiftDown() && me.
    isPrimaryButtonDown()) {
                cam.rz.setAngle(cam.rz.getAngle() - mouseDeltaX);
            }
            else if (me.isAltDown() && me.isPrimaryButtonDown()) {
                cam.ry.setAngle(cam.ry.getAngle() - mouseDeltaX);
                cam.rx.setAngle(cam.rx.getAngle() + mouseDeltaY);
            }
            else if (me.isAltDown() && me.isSecondaryButtonDown()) {
                double scale = cam.s.getX();
                double newScale = scale + mouseDeltaX * 0.01;
                cam.s.setX(newScale); cam.s.setY(newScale);
                cam.s.setZ(newScale);
            }
            else if (me.isAltDown() && me.isMiddleButtonDown()) {
                cam.t.setX(cam.t.getX() + mouseDeltaX);
                cam.t.setY(cam.t.getY() + mouseDeltaY);
            }
        }
    });
```

注意：轴心和角度定义了图像的目标点。指定轴心时，要仔细计算其值，否则图像可能会出现在不希望的地方。

3. 缩放比例

缩放变换会使节点看起来更大或更小，具体取决于缩放因子。缩放会更改节点，以便沿其轴的尺寸乘以比例因子。与旋转变换类似，缩放变换应用于轴心，该轴心被视为发生缩放的点。缩放可以使用缩放类和变换类中的缩放方法实现。在木琴应用程序中，可以在按住 Alt 键和鼠标右键的同时使用鼠标缩放木琴。缩放变换可用于查看缩放。Example 11-3 展示了比例变换的代码。

Example 11-3　Scaling

```
else if (me.isAltDown() && me.isSecondaryButtonDown()) {
    double scale = cam.s.getX();
    double newScale = scale + mouseDeltaX * 0.01;
```

第 11 章 基于 JavaFX 开发动画与视觉效果

```
        cam. s. setX (newScale); cam. s. setY (newScale); cam. s. setZ
(newScale);
}
...
```

4. 剪切

剪切变换旋转一个轴,使 x 轴和 y 轴不再垂直,节点的坐标按指定的乘数移动。剪切可以使用剪切类或变换类中的剪切方法实现。在木琴应用程序中,可以在按住 Shift 键的同时拖曳鼠标并按住鼠标左键以剪切木琴。Example 11-4 展示了剪切转换的代码片段。

Example 11-4　Scaling

```
else if (me.isShiftDown() && me.isPrimaryButtonDown()) {
    double yShear = shear.getY();
    shear.setY(yShear + mouseDeltaY/1000.0);
    double xShear = shear.getX();
    shear.setX(xShear + mouseDeltaX/1000.0);
}
```

5. 多重变换

通过指定一个有序的转换链可以构造多个转换转变。例如,可以缩放对象,然后对其应用剪切变换,也可以平移对象,然后对其进行缩放。Example 11-5 展示了应用于对象的多个变换,以创建木琴棒。

Example 11-5　Multiple Transformations

```
Cube base1Cube = new Cube(1.0, new Color(0.2, 0.12, 0.1, 1.0), 1.0);
base1Cube.setTranslateX(xStart + 135);
base1Cube.setTranslateZ(yPos+20.0);
base1Cube.setTranslateY(11.0);
base1Cube.setScaleX(barWidth * 11.5);
base1Cube.setScaleZ(10.0);
base1Cube.setScaleY(barDepth * 2.0);
```

Application FilesSource Code

- Xylophone.java
- NetBeans Projects

- transformations.zip

11.2 创建转换与时间轴动画

本节介绍在 JavaFX 中创建动画的知识,包括以下内容:动画基础——提供基本的动画概念——转换、时间轴动画、插补细分器。在示例应用中,以"树动画"为例描述树动画的实现,并提供一些有关 JavaFX 中实现动画应用的提示和技巧。

11.2.1 动画基础

JavaFX 中的动画可以分为时间轴动画和转换,下面将提供每种动画类型的示例,时间轴(Timeline)和转换(Transition)是 ation.Animation 的子类。

1. 转换

JavaFX 中提供了将动画合并到内部时间轴的方法,可以组合变换以创建并行或顺序执行的多个动画。有关详细信息,请参阅并行转换和顺序转换部分。以下内容将提供一些转换动画的示例。

2. 渐变转换

渐变转换会在给定时间内更改节点的不透明度。Example 11-6 展示了应用于矩形的淡入淡出转换的代码片段。首先创建一个圆角矩形,然后应用淡入转换。

Example 11-6 Fade Transition

```
final Rectangle rect1 = new Rectangle(10, 10, 100, 100);
rect1.setArcHeight(20);
rect1.setArcWidth(20);
rect1.setFill(Color.RED);
...
FadeTransition ft = new FadeTransition(Duration.millis(3000), rect1);
ft.setFromValue(1.0);
ft.setToValue(0.1);
ft.setCycleCount(Timeline.INDEFINITE);
ft.setAutoReverse(true);
ft.play();
```

3. 路径转换

路径转换在给定时间内沿路径将节点从一端移动到另一端。Example 11-7 展示了应用于矩形的路径转换的代码片段。当矩形到达路径末端时,动画将反转。在代码中,首先创建一个圆角矩形,然后创建一个新的路径动画并应用于该矩形。

Example 11-7　Path Transition

```
final Rectangle rectPath = new Rectangle (0, 0, 40, 40);
rectPath.setArcHeight(10);
rectPath.setArcWidth(10);
rectPath.setFill(Color.ORANGE);
...
Path path = new Path( );
path.getElements( ).add(new MoveTo(20,20));
path.getElements( ).add(new CubicCurveTo(380, 0, 380, 120, 200, 120));
path.getElements( ).add(new CubicCurveTo(0, 120, 0, 240, 380, 240));
PathTransition pathTransition = new PathTransition( );
pathTransition.setDuration(Duration.millis(4000));
pathTransition.setPath(path);
pathTransition.setNode(rectPath);
pathTransition. setOrientation ( PathTransition. OrientationType.
ORTHOGONAL_TO_TANGENT);
pathTransition.setCycleCount(Timeline.INDEFINITE);
pathTransition.setAutoReverse(true);
pathTransition.play( );
```

4. 并行转换

并行转换可以同时执行多个转换。Example 11-8 展示了并行转换的代码片段,该并行转换执行应用于矩形的淡入、平移、旋转和缩放转换。

Example 11-8　Parallel Transition

```
Rectangle rectParallel = new Rectangle(10,200,50, 50);
rectParallel.setArcHeight(15);
rectParallel.setArcWidth(15);
rectParallel.setFill(Color.DARKBLUE);
rectParallel.setTranslateX(50);
rectParallel.setTranslateY(75);
```

```
...
FadeTransition fadeTransition = new FadeTransition(Duration.millis
(3000), rectParallel);
fadeTransition.setFromValue(1.0f);
fadeTransition.setToValue(0.3f);
fadeTransition.setCycleCount(2);
fadeTransition.setAutoReverse(true);
TranslateTransition translateTransition = new TranslateTransition
(Duration.millis(2000), rectParallel);
translateTransition.setFromX(50);
translateTransition.setToX(350);
translateTransition.setCycleCount(2);
translateTransition.setAutoReverse(true);
RotateTransition rotateTransition = new RotateTransition(Duration.
millis(3000), rectParallel);
rotateTransition.setByAngle(180f);
rotateTransition.setCycleCount(4);
rotateTransition.setAutoReverse(true);
ScaleTransition scaleTransition = new ScaleTransition(Duration.
millis(2000), rectParallel);
scaleTransition.setToX(2f);
scaleTransition.setToY(2f);
scaleTransition.setCycleCount(2);
scaleTransition.setAutoReverse(true);
parallelTransition = new ParallelTransition();
parallelTransition. getChildren ( ). addAll ( fadeTransition,
translateTransition, rotateTransition, scaleTransition);
parallelTransition.setCycleCount(Timeline.INDEFINITE);
parallelTransition.play();
```

5. 顺序转换

顺序转换可以一个接一个地执行多个转换。Example 11-9 展示了一个接一个地执行的顺序转换的代码片段，应用于矩形的淡入、平移、旋转和缩放变换。

Example 11-9　Sequential Transition

```
Rectangle rectSeq = new Rectangle(25,25,50,50);
rectSeq.setArcHeight(15);
rectSeq.setArcWidth(15);
```

```
rectSeq.setFill(Color.CRIMSON);
rectSeq.setTranslateX(50);
rectSeq.setTranslateY(50);
...
FadeTransition fadeTransition = new FadeTransition(Duration.millis
(1000), rectSeq);
fadeTransition.setFromValue(1.0f);
fadeTransition.setToValue(0.3f);
fadeTransition.setCycleCount(1);
fadeTransition.setAutoReverse(true);
TranslateTransition translateTransition = new TranslateTransition
(Duration.millis(2000), rectSeq);
translateTransition.setFromX(50);
translateTransition.setToX(375);
translateTransition.setCycleCount(1);
translateTransition.setAutoReverse(true);
RotateTransition rotateTransition = new RotateTransition(Duration.
millis(2000), rectSeq);
rotateTransition.setByAngle(180f);
rotateTransition.setCycleCount(4);
rotateTransition.setAutoReverse(true);
ScaleTransition scaleTransition = new ScaleTransition (Duration.
millis(2000), rectSeq);
scaleTransition.setFromX(1);
scaleTransition.setFromY(1);
scaleTransition.setToX(2);
scaleTransition.setToY(2);
scaleTransition.setCycleCount(1);
scaleTransition.setAutoReverse(true);
sequentialTransition = new SequentialTransition();
sequentialTransition. getChildren ( ). addAll ( fadeTransition,
translateTransition, rotateTransition, scaleTransition);
sequentialTransition.setCycleCount(Timeline.INDEFINITE);
sequentialTransition.setAutoReverse(true);
sequentialTransition.play();
```

有关动画和转换的更多信息,请参阅 SDK 的集成项目中的 API 文档和动画部分。

11.2.2 时间轴动画

动画由其相关属性驱动,例如大小、位置和颜色等。时间轴提供了沿时间进程更新属性值的功能。JavaFX 支持关键帧动画,在关键帧动画中,图形场景的动画状态转换通过特定时间场景状态的开始和结束快照(关键帧)进行声明。系统可以自动执行动画,并可以在需要时停止、暂停、恢复、反转或重复移动。

1. 基本时间轴动画

Example 11-10 中的代码的作用是水平设置矩形动画,并在 500ms 内将其从原始位置 X=100 移动到 X=300。要水平设置对象动画,需要更改 X 坐标并保持 Y 坐标不变。

Example 11-10　Timeline Animation

```
final Rectangle rectBasicTimeline = new Rectangle(100, 50, 100, 50);
rectBasicTimeline.setFill(Color.RED);
...
final Timeline timeline = new Timeline();
timeline.setCycleCount(Timeline.INDEFINITE);
timeline.setAutoReverse(true);
final KeyValue kv = new KeyValue(rectBasicTimeline.xProperty(), 300);
final KeyFrame kf = new KeyFrame(Duration.millis(500), kv);
timeline.getKeyFrames().add(kf);
timeline.play();
```

2. 基本时间轴事件

JavaFX 提供了将时间线播放期间可以触发的事件合并到一起的方法。Example 11-11 中的代码在指定范围内更改了圆的半径,并且关键帧触发了圆在场景 X 坐标中的随机变换。

Example 11-11　Timeline Events

```
import javafx.application.Application;
import javafx.stage.Stage;
import javafx.animation.AnimationTimer;
import javafx.animation.KeyFrame;
import javafx.animation.KeyValue;
import javafx.animation.Timeline;
```

```java
import javafx.event.ActionEvent;
import javafx.event.EventHandler;
import javafx.scene.Group;
import javafx.scene.Scene;
import javafx.scene.effect.Lighting;
import javafx.scene.layout.StackPane;
import javafx.scene.paint.Color;
import javafx.scene.shape.Circle;
import javafx.scene.text.Text;
import javafx.util.Duration;
public class TimelineEvents extends Application {
    private Timeline timeline;          //main timeline
    private AnimationTimer timer;
    //用于存储实际帧的变量
    private Integer i=0;
    @Override public void start(Stage stage) {
        Group p = new Group();
        Scene scene = new Scene(p);
        stage.setScene(scene);
        stage.setWidth(500);
        stage.setHeight(500);
        p.setTranslateX(80);
        p.setTranslateY(80);
        final Circle circle = new Circle(20, Color.rgb(156,216,255));
                                    //create a circle with effect
        circle.setEffect(new Lighting());
        final Text text = new Text (i.toString());
                                    //create a text inside a circle
        text.setStroke(Color.BLACK);
        //创建包含文字的圆形布局
        final StackPane stack = new StackPane();
        stack.getChildren().addAll(circle, text);
        stack.setLayoutX(30);
        stack.setLayoutY(30);
        p.getChildren().add(stack);
        stage.show();
        //创建用于移动圆的时间线
        timeline = new Timeline();
        timeline.setCycleCount(Timeline.INDEFINITE);
```

```
            timeline.setAutoReverse(true);
         timer = new AnimationTimer() {//You can add a specific action when each frame is started.
            @Override
            public void handle(long l) {
                text.setText(i.toString());
                i++;
            }
        };
        //使用factory创建keyValue:缩放圆两次
        KeyValue keyValueX = new KeyValue(stack.scaleXProperty(), 2);
        KeyValue keyValueY = new KeyValue(stack.scaleYProperty(), 2);
        //创建关键帧,在2s内达到keyValue
        Duration duration = Duration.millis(2000);
        //可以在到达关键帧时添加特定事件处理
        EventHandler onFinished = new EventHandler<ActionEvent>() {
            public void handle(ActionEvent t) {
                stack.setTranslateX(java.lang.Math.random() * 200-100);
                //重置计数器
                i = 0;
            }
        };
         KeyFrame keyFrame = new KeyFrame(duration, onFinished, keyValueX, keyValueY);
            timeline.getKeyFrames().add(keyFrame);//Add the keyframe to the timeline
        timeline.play();
        timer.start();
    }
    public static void main(String[] args) {
        Application.launch(args);
    }
}
```

运行结果如图11.2所示。

3. 插值器

插值定义了对象在运动起点和终点之间的位置,可以使用插值器类的各种

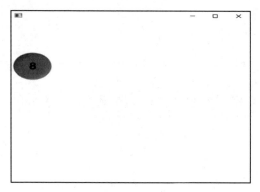

图 11.2　TimelineEvents 的运行结果

内置实现,也可以通过自己的插值器实现自定义插值的行为。

4. 内置插值器

JavaFX 提供了几个内置插值器,可用于在动画中创建不同的效果。默认情况下,JavaFX 使用线性插值计算坐标。Example 11-12 展示了一个代码片段,其中,EASE_Duth interpolator 对象实例被添加到基本时间线动画的 KeyValue 中。当对象到达其起点和终点时,该插值器会创建弹簧效果。

Example 11-12　Built-in Interpolator

```
final Rectangle rectBasicTimeline = new Rectangle(100, 50, 100, 50);
rectBasicTimeline.setFill(Color.BROWN);
...
final Timeline timeline = new Timeline();
timeline.setCycleCount(Timeline.INDEFINITE);
timeline.setAutoReverse(true);
final KeyValue kv = new KeyValue(rectBasicTimeline.xProperty(),
300,
Interpolator.EASE_BOTH);
final KeyFrame kf = new KeyFrame(Duration.millis(500), kv);
timeline.getKeyFrames().add(kf);
timeline.play();
```

5. 自定义插值器

除了内置的插值器,还可以通过自己的插值器实现自定义插值的行为。自定义插值器示例由两个 Java 文件组成。Example 11-13 展示了用于计算动画 Y 坐标的自定义插值器。Example 11-14 展示了使用 AnimationBooleanInterpolator 的动画的代码片段。

Example 11-13　Custom Interpolator

```
public class AnimationBooleanInterpolator extends Interpolator {
    @Override
    protected double curve(double t) {
        return Math.abs(0.5-t) * 2 ;
    }
}
```

Example 11-14　Animation with Custom Interpolator

```
final KeyValue keyValue1 = new KeyValue(rect.xProperty( ), 300);
AnimationBooleanInterpolator yInterp = new
AnimationBooleanInterpolator( );
final KeyValue keyValue2 = new KeyValue (rect. yProperty ( ), 0.,
yInterp);
```

Application Files
NetBeans Projects
animations.zip

11.2.3　树动画示例

本节提供有关树动画示例的详细信息，下面介绍场景中的所有元素是如何创建和设置动画的。图 11.3 所示为带有树的场景的示意图。

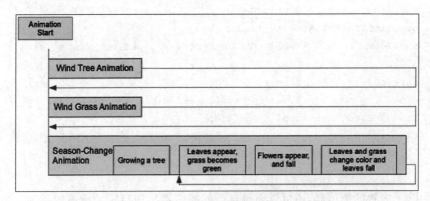

图 11.3　动画时间轴

1. 项目和要素

树动画项目由几个文件组成。每个元素（如树叶、草叶等）都是在单独的类中创建的。TreeGenerator 类从所有元素中创建一棵树。Animator 类包含除 GrassWindAnimation 类中的 grass 动画之外的所有动画。示例中的场景包含以下元素：

- 有树枝、叶子和花的树；
- 草。

每个元素都以自己的方式设置动画，有些动画并行运行，有些动画按顺序运行。树木生长动画仅运行一次，而季节变化动画则设置为无限运行。

季节变化动画包括以下部分：

- 树叶和花朵出现在树上；
- 花瓣落下并消失；
- 树叶和草会变色；
- 树叶落在地上并消失。

2. 创建草地

在树动画示例中，草（如图 11-3 所示）由单独的草叶组成，每个草叶都使用路径创建并添加到列表中，然后每个叶片都会弯曲并着色。使用一种算法随机化叶片的高度、曲线和颜色，并将叶片分布在"地面"上，可以指定叶片的数量和被草覆盖的"地面"的大小，如 Example 11-5 所示。

Example 11-15　Creating a Grass Blade

```
public class Blade extends Path {
    public final Color SPRING_COLOR = Color.color(random() * 0.5,
random() * 0.5 + 0.5, 0.).darker();
    public final Color AUTUMN_COLOR = Color.color(random() * 0.4 + 0.
3, random() * 0.1 + 0.4, random() * 0.2);
    private final static double width = 3;
    private double x = RandomUtil.getRandom(170);
    private double y = RandomUtil.getRandom(20) + 20;
    private double h = (50 * 1.5 - y / 2) * RandomUtil.getRandom(0.3);
    public SimpleDoubleProperty phase = new SimpleDoubleProperty();
    public Blade() {
        getElements().add(new MoveTo(0, 0));
        final QuadCurveTo curve1;
        final QuadCurveTo curve2;
```

```
        getElements().add(curve1 = new QuadCurveTo(-10, h, h / 4, h));
        getElements().add(curve2 = new QuadCurveTo(-10, h, width, 0));
        setFill(AUTUMN_COLOR); //autumn color of blade
        setStroke(null);
        getTransforms().addAll(Transform.translate(x, y));
        curve1.yProperty().bind(new DoubleBinding() {
            super.bind(curve1.xProperty());
        }
        @Override
        protected double computeValue() {
          final double xx0 = curve1.xProperty().get();
          return Math.sqrt(h * h - xx0 * xx0);
        }
    });                                                    //叶片顶部路径为圆形
//树叶倾斜的代码
    curve1.controlYProperty().bind(curve1.yProperty().add(-h / 4));
    curve2.controlYProperty().bind(curve1.yProperty().add(-h / 4));
    curve1.xProperty().bind(new DoubleBinding() {
      final double rand = RandomUtil.getRandom(PI / 4);
      {  super.bind(phase);   }
      @Override
      protected double computeValue() {
        return (h / 4) + ((cos(phase.get() + (x + 400.) * PI / 1600 +
rand) + 1) / 2.) * (-3./ 4) * h;
      }
    });
  }
}
```

3. 为草地运动创建时间轴动画

更改叶片顶部 X 坐标的时间轴动画以创建草运动。使用几种算法使运动看起来更自然，例如，每个叶片的顶部都以圆形而非直线移动，叶片的侧曲线使叶片看起来像是在风下弯曲。添加随机数以分隔每个叶片的移动，如 Example 11-16 所示。

Example 11-16 Grass Animation

```
class GrassWindAnimation extends Transition {
    final private Duration animationTime = Duration.seconds(3);
```

```
final private DoubleProperty phase = new SimpleDoubleProperty(0);
final private Timeline tl = new Timeline(Animation.INDEFINITE);
public GrassWindAnimation(List<Blade> blades) {
    setCycleCount(Animation.INDEFINITE);
    setInterpolator(Interpolator.LINEAR);
    setCycleDuration(animationTime);
    for (Blade blade : blades) {
      blade.phase.bind(phase);
    }
}
@Override
protected void interpolate(double frac) {
    phase.set(frac * 2 * PI);
}
}
```

4. 树

一棵树由树枝、叶子和花组成,叶子和花被画在树的顶端树枝上,每个分支生成由从父代分支绘制的 3 个分支(一个顶部分支和两个侧面分支)组成。可以使用在主类的 TreeGenerator 构造方法中传递的"_分支_代数"指定代码中的"代数"。Example 11-17 展示了 TreeGenerator 类,它创建了树的主干(或根分支),并为下一代添加了 3 个分支。

Example 11-17 Root Branch

```
private List < Branch > generateBranches (Branch parentBranch, int depth) {
    List<Branch> branches = new ArrayList<>();
    if (parentBranch == null) {                    //添加根分支
      branches.add(new Branch());
    } else {
        if (parentBranch.length < 10) {
          return Collections.emptyList();
        }
      branches.add(new Branch(parentBranch, Type.LEFT, depth));
      branches.add(new Branch(parentBranch, Type.RIGHT, depth));
      branches.add(new Branch(parentBranch, Type.TOP, depth));
    }
```

```
        return branches;
    }
```

为了使树看起来更自然,每个子代分支都与父代分支呈一定角度生长,并且每个子代分支都比其父代分支小。子代分支的角度使用随机值计算。Example 11-18 展示了创建子分支的代码。

Example 11-18   Child Branches

```
public Branch(Branch parentBranch, Type type, int depth) {
    this();
    SimpleDoubleProperty locAngle = new SimpleDoubleProperty(0);
    globalAngle.bind(locAngle.add(parentBranch.globalAngle.get()));
    double transY = 0;
    switch (type) {
      case TOP:
        transY = parentBranch.length;
        length = parentBranch.length * 0.8;
        locAngle.set(getRandom(10));
        break;
      case LEFT:
      case RIGHT:
       transY = parentBranch.length - getGaussianRandom(0, parentBranch.length, parentBranch.length / 10, parentBranch.length / 10);
        locAngle.set(getGaussianRandom(35, 10) * (Type.LEFT == type ? 1 : -1));
        if ((0 > globalAngle.get() || globalAngle.get() > 180) && depth < 4) {
            length = parentBranch.length * getGaussianRandom(0.3, 0.1);
        } else {
            length = parentBranch.length * 0.6;
        }
        break;
    }
    setTranslateY(transY);
    getTransforms().add(new Rotate(locAngle.get(), 0, 0));
    globalH = getTranslateY() * cos(PI /2 - parentBranch.globalAngle.get() * PI / 180) + arentBranch.globalH;
```

```
        setBranchStyle(depth);
        addChildToParent(parentBranch, this);
    }
```

5. 树叶和花朵

叶子是在上面的树枝上形成的。因为叶子与树的分支是同时创建的,所以叶子按叶子缩放为 0。如 Example 11-19 所示,设置 scaley(0)以在树生长之前隐藏它们,同样的技巧也被用来在树叶落下时隐藏它们。为了营造更自然的外观,树叶的绿色色调略有不同。此外,叶子的颜色也会随着叶子的位置而变化,例如较暗的色调将应用于树冠中部下方的叶子。

Example 11-19　Leaf Shape and Placement

```
public class Leaf extends Ellipse {
    public final Color AUTUMN_COLOR;
    private final int N = 5;
    private List<Ellipse> petals = new ArrayList<>(2 * N + 1);
    public Leaf(Branch parentBranch) {
        super(0, parentBranch.length / 2., 2, parentBranch.length / 2.);
        setScaleX(0);
        setScaleY(0);
        double rand = random() * 0.5 + 0.3;
        AUTUMN_COLOR = Color.color(random() * 0.1 + 0.8, rand, rand / 2);
        Color color = new Color(random() * 0.5, random() * 0.5 + 0.5, 0, 1);
        if (parentBranch.globalH < 400 && random() < 0.8) {
                                                           //bottom leaf is
          darker
            color = color.darker();
        }
        setFill(color);
    }
}
```

花卉在 Flower 类中创建,然后添加到 TreeGenerator 类中树的顶部分支。可以指定花的花瓣数。花瓣是一些重叠的椭圆,呈圆形分布。与草和叶子类似,花瓣的颜色也有不同的粉红色。

6. 为树元素设置动画

本节介绍树动画示例中设置树和季节变化的动画技术。平行过渡用于启动场景中的所有动画,如 Example 11-20 所示。

Example 11-20　Main Animation

```
final Transition all = new ParallelTransition(new GrassWindAnimation
(grass), treeWindAnimation, new SequentialTransition
(branchGrowingAnimation, seasonsAnimation(tree, grass)));
all.play();
```

7. 种树

动画树只在开始时生长一次。应用程序启动了一个连续的过渡动画,一代又一代地增长分支,如 Example 11-21 所示。最初,长度设置为 0,根分支的大小和角度在 TreeGenerator 类中指定。目前,每一代都是在两秒内成长起来的。

Example 11-21　Sequential Transition to Start Branch Growing Animation

```
SequentialTransition branchGrowingAnimation = new
SequentialTransition();
```

Example 11-22 中的代码创建了树生长动画。

Example 11-22　Branch Growing Animation

```
private Animation animateBranchGrowing ( List < Branch >
branchGeneration) {
    ParallelTransition sameDepthBranchAnimation = new
ParallelTransition();
    for (final Branch branch : branchGeneration) {
        Timeline branchGrowingAnimation = new Timeline(new KeyFrame
(duration, new KeyValue ( branch. base. endYProperty ( ), branch.
length)));
        PauseTransition pauseTransition = new PauseTransition();
        pauseTransition. setOnFinished ( t - > branch. base.
setStrokeWidth(branch.length / 25));
            sameDepthBranchAnimation. getChildren ( ). add ( new
SequentialTransition(pauseTransition,
branchGrowingAnimation));
    }
```

```
        return sameDepthBranchAnimation;
}
```

因为所有的分支线都是同时计算和创建的,所以它们可以作为点出现在场景中。这段代码引入了一些技巧以在代码增长之前隐藏代码行,例如代码持续时间,1毫秒会使过渡暂停一段不明显的时间。在 Example 11-23 中,base.setStrokeWidth(0)代码在每一代的生长动画开始之前将分支宽度设置为 0。

Example 11-23　Tree Growing Animation Optimization

```
private void setBranchStyle(int depth) {
    base.setStroke(Color.color(0.4, 0.1, 0.1, 1));
    if (depth < 5) {
        base.setStrokeLineJoin(StrokeLineJoin.ROUND);
        base.setStrokeLineCap(StrokeLineCap.ROUND);
    }
    base.setStrokeWidth(0);
}
```

8. 创建树冠运动

在树木生长的同时,风动画开始,树枝、树叶和花朵一起移动。风动画类似于草运动动画,但其更简单,因为只有树枝的角度会发生变化。为了使树木的运动看起来更加自然,弯曲角度对于不同的树枝世代而言是不同的。分支的生成量越高(分支越小),其弯曲程度就越大。Example 11-24 提供了风动画的代码。

Example 11-24　Wind Animation

```
private Animation animateTreeWind(List<Branch> branchGeneration, int depth) {
    ParallelTransition wind = new ParallelTransition();
    for (final Branch brunch : branchGeneration) {
        final Rotate rotation = new Rotate(0);
        brunch.getTransforms().add(rotation);
        Timeline windTimeline = new Timeline(new KeyFrame(WIND_CYCLE_DURATION, new KeyValue(rotation.angleProperty(), depth * 2)));
        windTimeline.setAutoReverse(true);
        windTimeline.setCycleCount(Animation.INDEFINITE);
        wind.getChildren().add(windTimeline);
    }
```

```
        return wind;
    }
```

9. 让季节变化充满活力

季节变化动画实际上是在树生长之后开始的,并且无限期地运行。Example 11-25 中的代码调用了所有的季节动画。

Example 11-25　Starting Season Animation

```
private Transition seasonsAnimation(final Tree tree, final List<Blade> grass) {
    Transition spring = animateSpring(tree.leafage, grass);
    Transition flowers = animateFlowers(tree.flowers);
    Transition autumn = animateAutumn(tree.leafage, grass);
    SequentialTransition sequentialTransition = new SequentialTransition(spring, flowers, autumn);
    return sequentialTransition;
}
private Transition animateSpring(List<Leaf> leafage, List<Blade> grass) {
    ParallelTransition springAnimation = new ParallelTransition();
    for (final Blade blade : grass) {
        springAnimation.getChildren().add(new FillTransition(GRASS_BECOME_GREEN_DURATION,
blade, (Color) blade.getFill(), blade.SPRING_COLOR));
    }
    for (Leaf leaf : leafage) {
        ScaleTransition leafageAppear = new ScaleTransition(LEAF_APPEARING_DURATION, leaf);
        leafageAppear.setToX(1);
        leafageAppear.setToY(1);
        springAnimation.getChildren().add(leafageAppear);
    }
    return springAnimation;
}
```

如 Example 11-26 所示,一旦所有树枝都长出来了,叶子就开始按照图中的指示出现。

Example 11-26　Parallel Transition to Start Spring Animation and Show Leaves

```
private Transition animateSpring(List<Leaf> leafage, List<Blade> grass) {
    ParallelTransition springAnimation = new ParallelTransition();
    for (final Blade blade : grass) {
        springAnimation.getChildren().add(new FillTransition(GRASS_BECOME_GREEN_DURATION, blade, (Color) blade.getFill(), blade.SPRING_COLOR));
    }
    for (Leaf leaf : leafage) {
        ScaleTransition leafageAppear = new ScaleTransition(LEAF_APPEARING_DURATION, leaf);
        leafageAppear.setToX(1);
        leafageAppear.setToY(1);
        springAnimation.getChildren().add(leafageAppear);
    }
    return springAnimation;
}
```

当所有叶子都可见时，花朵开始出现，如 Example 11-27 所示。顺序过渡用于逐渐显示花朵。在 Example 11-27 的顺序转换代码中设置花朵外观的延迟，花只出现在树冠上。

Example 11-27　Showing Flowers

```
private Transition animateFlowers(List<Flower> flowers) {
    ParallelTransition flowersAppearAndFallDown = new ParallelTransition();
    for (int i = 0; i < flowers.size(); i++) {
        final Flower flower = flowers.get(i);
        for (Ellipse pental : flower.getPetals()) {
            FadeTransition flowerAppear = new FadeTransition(FLOWER_APPEARING_DURATION, petal);
            flowerAppear.setToValue(1);
            flowerAppear.setDelay(FLOWER_APPEARING_DURATION.divide(3).multiply(i + 1));
```

```
            flowersAppearAndFallDown.getChildren().add(new
SequentialTransition ( new  SequentialTransition  ( flowerAppear,
fakeFallDownAnimation(petal))));
        }
    }
    return flowersAppearAndFallDown;
}
```

一旦所有的花都出现在屏幕上,它们的花瓣就开始掉落。在 Example 11-28 中,花被复制,第一组花被隐藏,以便稍后显示。

Example 11-28   Duplicating Petals

```
private Ellipse copyEllipse(Ellipse petalOld, Color color) {
    Ellipse ellipse = new Ellipse();
    ellipse.setRadiusX(petalOld.getRadiusX());
    ellipse.setRadiusY(petalOld.getRadiusY());
    if (color == null) {
        ellipse.setFill(petalOld.getFill());
    } else {
        ellipse.setFill(color);
    }
    ellipse.setRotate(petalOld.getRotate());
    ellipse.setOpacity(0);
    return ellipse;
}
```

复制的花瓣开始一个接一个地落在地上,如 Example 11-29 所示。花瓣落在地上 5 秒后就消失了。花瓣的下落轨迹不是直线,而是经过计算的正弦曲线,因此花瓣下落时似乎是在旋转。

Example 11-29   Shedding Flowers

```
Animation  fakeLeafageDown  =  fakeFallDownEllipseAnimation ( leaf,
leaf.AUTUMN_COLOR, node -> {
    node.setScaleX(0);
    node.setScaleY(0);
});
```

当所有的花都从场景中消失时,下一个季节就开始了。树叶和草都变黄了,树叶落下来并消失了。Example 11-30 中使用的花瓣掉落的相同算法用于显示落叶,并启动秋季动画。

Example 11-30　Animating Autumn Changes

```
private Transition animateAutumn(List<Leaf> leafage, List<Blade>
grass) {
    ParallelTransition autumn = new ParallelTransition();
    ParallelTransition yellowLeafage = new ParallelTransition();
    ParallelTransition dissappearLeafage = new ParallelTransition();
    for (final Leaf leaf : leafage) {
        final FillTransition toYellow = new FillTransition(LEAF_
BECOME_YELLOW_DURATION, leaf, null, leaf.AUTUMN_COLOR);
        Animation fakeLeafageDown = fakeFallDownEllipseAnimation
(leaf,leaf.AUTUMN_COLOR,node -> {
            node.setScaleX(0);
            node.setScaleY(0);
        });
        dissappearLeafage.getChildren().add(fakeLeafageDown);
    }
    ParallelTransition grassBecomeYellowAnimation = new
ParallelTransition();
    for (final Blade blade : grass) {
        final FillTransition toYellow = new FillTransition(GRASS_
BECOME_ YELLOW_ DURATION, blade, (Color) blade.getFill( ), blade.
AUTUMN_COLOR);
        toYellow.setDelay(Duration.seconds(1 * random()));
        grassBecomeYellowAnimation.getChildren().add(toYellow);
    }
    autumn.getChildren().addAll(grassBecomeYellowAnimation, new
SequentialTransition(yellowLeafage, dissappearLeafage));
    return autumn;
}
```

当所有的叶子都从地上消失后,春天动画开始将草染成绿色并显示叶子。这个JavaFX动画应用的运行结果如图11.4所示。

Application Files

- NetBeans Projects
- tree_animation.zip

图 11.4 树动画的运行结果

11.3 创建视觉效果

本节包含以下 JavaFX 应用的开发——混合效果、布鲁姆效应、模糊效果、阴影效果、内阴影效应、反射灯光效果、透视效果、创造连锁反应。

11.3.1 应用效果

本节将介绍如何使用视觉效果增强 JavaFX 应用的外观。所有的效果都位于 javafx.scene.effect 包以及 Effect 类的子类中。

1. 混合效果

混合(blend)是一种使用预定义的混合模式之一将两个输入组合在一起的效果。对于节点混合(node.setBlendMode( )),两个输入为:

- 正在渲染的节点(顶部输入);
- 节点下的所有内容(底部输入)。

底部输入的确定基于以下规则:

- 所有较低级别的兄弟姐妹都包括在同一组中;
- 如果组具有定义的混合模式,则过程停止,并定义底部输入;
- 如果组具有默认的混合模式,则使用相同的规则递归地包含组下的所有内容;
- 如果进程递归地返回到根节点,则包含场景的背景绘制。

> **注意**:如果底部输入包含场景的背景绘制(通常为不透明颜色),则 SRC_Top 模式在完全不透明的底部源上渲染,并且没有效果,在这种情况下,SRC_Top 模式相当于 SRC_OVER 模式。

混合模式定义了对象混合在一起的方式。例如,在图 11.5 中,可以看到一些应用于一个用正方形分组的圆的混合模式的示例。

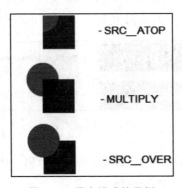

图 11.5 混合模式的示例

Example 11-31 展示了示例应用程序中混合效果的代码片段。

Example 11-31 Blend Effect

```
static Node blendMode() {
    Rectangle r = new Rectangle();
    r.setX(590);
    r.setY(50);
    r.setWidth(50);
    r.setHeight(50);
    r.setFill(Color.BLUE);
    Circle c = new Circle();
    c.setFill(Color.RED);
    c.setCenterX(590);
    c.setCenterY(50);
    c.setRadius(25);
    c.setBlendMode(BlendMode.SRC_ATOP);
    Group g = new Group();
    g.setBlendMode(BlendMode.SRC_OVER);
    g.getChildren().add(r);
    g.getChildren().add(c);
    return g;
}
```

2. Bloom 效果

Bloom 效果可以使图像中较亮的部分看起来发光,这基于可配置的阈值,阈值从 0.0 到 1.0 不等。默认情况下,阈值为 0.3。

图 11.6 所示为默认阈值和阈值为 1.0 时的 Bloom 效应。

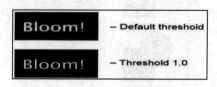

图 11.6 Bloom 效果示例

Example 11-32 展示了使用 Bloom 效果的示例应用程序的代码片段。

Example 11-32 Bloom Example

```
static Node bloom() {
    Group g = new Group();
```

```
Rectangle r = new Rectangle( );
r.setX(10);
r.setY(10);
r.setWidth(160);
r.setHeight(80);
r.setFill(Color.DARKBLUE);
Text t = new Text( );
t.setText("Bloom!");
t.setFill(Color.YELLOW);
t.setFont(Font.font("null", FontWeight.BOLD, 36));
t.setX(25);
t.setY(65);
g.setCache(true);
//g.setEffect(new Bloom( ));
Bloom bloom = new Bloom( );
bloom.setThreshold(1.0);
g.setEffect(bloom);
g.getChildren( ).add(r);
g.getChildren( ).add(t);
g.setTranslateX(350);
return g;
}
```

3. 模糊效果

模糊(Blur)是常见的效果,可用于为选定对象提供更多的焦点。使用 JavaFX 可以应用 BoxBlur、运动模糊(MotionBlur)或高斯模糊(GaussianBlur)。

1) BoxBlur

BoxBlur 是一种模糊效果,它使用一个简单的框过滤器内核,单独使用两个维度中的可配置大小控制应用于对象的模糊量,以及控制生成模糊质量的迭代参数。图 11.7 所示为两个模糊文本示例。

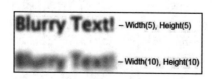

图 11.7 模糊文本示例

Example 11-33 是一个使用 BoxBlur 效果的代码片段。

Example 11-33　BoxBlur Example

```
static Node boxBlur( ) {
    Text t = new Text( );
    t.setText("Blurry Text!");
    t.setFill(Color.RED);
    t.setFont(Font.font("null", FontWeight.BOLD, 36));
    t.setX(10);
    t.setY(40);
    BoxBlur bb = new BoxBlur( );
    bb.setWidth(5);
    bb.setHeight(5);
    bb.setIterations(3);
    t.setEffect(bb);
    t.setTranslateX(300);
    t.setTranslateY(100);
    return t;
}
```

2）运动模糊

运动模糊效果使用可配置半径和角度的高斯模糊创建移动对象的效果。图 11.8 所示为运动模糊对文本的影响。

Example 11-34 展示了在示例应用程序中创建运动模糊效果的代码片段，该效果的半径设置为 15，角度设置为 45。

Example 11-34　Motion Blur Example

```
static Node motionBlur( ) {
    Text t = new Text( );
    t.setX(20.0f);
    t.setY(80.0f);
    t.setText("Motion Blur");
    t.setFill(Color.RED);
    t.setFont(Font.font("null", FontWeight.BOLD, 60));
    MotionBlur mb = new MotionBlur();
    mb.setRadius(15.0f);
    mb.setAngle(45.0f);
    t.setEffect(mb);
    t.setTranslateX(300);
```

```
        t.setTranslateY(150);
        return t;
}
```

3）高斯模糊

高斯模糊是使用可配置半径的高斯算法模糊对象的效果。图 11.9 所示为高斯模糊对文本的影响。Example 11-35 展示了一个使用高斯模糊效果模糊文本的代码片段。

图 11.8　运动模糊示例　　　　图 11.9　高斯模糊示例

Example 11-35　Gaussian Blur

```
static Node gaussianBlur( ) {
    Text t2 = new Text( );
    t2.setX(10.0f);
    t2.setY(140.0f);
    t2.setCache(true);
    t2.setText("Gaussian Blur");
    t2.setFill(Color.RED);
    t2.setFont(Font.font("null", FontWeight.BOLD, 36));
    t2.setEffect(new GaussianBlur());
    return t2;
}
```

4．阴影效果

阴影效果用于渲染内容的阴影，可以指定阴影的颜色、半径、偏移和其他参数。图 11.10 所示为不同对象上的阴影效果。

图 11.10　阴影效果示例

Example 11-36 展示了如何在文本和圆上创建阴影。

Example 11-36　Text and Circle With Shadows

```
import javafx.collections.ObservableList;
import javafx.application.Application;
import javafx.scene.*;
import javafx.stage.*;
import javafx.scene.shape.*;
import javafx.scene.effect.*;
import javafx.scene.paint.*;
import javafx.scene.text.*;
public class HelloEffects extends Application {
    Stage stage;
    Scene scene;
    @Override
    public void start(Stage stage) {
        stage.show();
        scene = new Scene(new Group(), 840, 680);
         ObservableList<Node> content = ((Group) scene.getRoot()).getChildren();
        content.add(dropShadow());
        stage.setScene(scene);
    }
    static Node dropShadow() {
        Group g = new Group();
        DropShadow ds = new DropShadow();
        ds.setOffsetY(3.0);
        ds.setOffsetX(3.0);
        ds.setColor(Color.GRAY);
        Text t = new Text();
        t.setEffect(ds);
        t.setCache(true);
        t.setX(20.0f);
        t.setY(70.0f);
        t.setFill(Color.RED);
        t.setText("JavaFX drop shadow effect");
        t.setFont(Font.font("null", FontWeight.BOLD, 32));
        DropShadow ds1 = new DropShadow();
        ds1.setOffsetY(4.0f);
        ds1.setOffsetX(4.0f);
```

```
        ds1.setColor(Color.CORAL);
        Circle c = new Circle();
        c.setEffect(ds1);
        c.setCenterX(50.0f);
        c.setCenterY(325.0f);
        c.setRadius(30.0f);
        c.setFill(Color.RED);
        c.setCache(true);
        g.getChildren().add(t);
        g.getChildren().add(c);
        return g;
    }
    public static void main(String[] args) {
        Application.launch(args);
    }
}
```

提示:
- 阴影过宽会使元素看起来沉重,阴影颜色应该是真实的,通常是比背景色更暗的色调;
- 如果有多个具有阴影效果的对象,确定放置方向可以对所有对象使用相同的阴影,投下的阴影会在具有从一个方向射出的光的外观物体上留下阴影。

11.3.2 内部阴影效果

内部阴影效果是在给定内容的边缘以指定的颜色、半径和偏移渲染阴影的效果。图 11.11 所示为纯文本和应用内部阴影效果的相同文本。Example 11-37 展示了如何在文本上创建内部阴影。

图 11.11 内部阴影效果示例

Example 11-37 Inner Shadow

```
static Node innerShadow() {
```

```
        InnerShadow is = new InnerShadow( );
        is.setOffsetX(2.0f);
        is.setOffsetY(2.0f);
        Text t = new Text( );
        t.setEffect(is);
        t.setX(20);
        t.setY(100);
        t.setText("Inner Shadow");
        t.setFill(Color.RED);
        t.setFont(Font.font("null", FontWeight.BOLD, 80));
        t.setTranslateX(300);
        t.setTranslateY(300);
        return t;
    }
```

11.3.3 反射效果

反射是在实际对象下方渲染对象的反射的效果。注意：具有反射效果的节点反射不会响应鼠标事件或节点上包含的方法。

图 11.12 所示为应用于文本的反射效果。使用 setFraction() 方法指定可见反射的量。Example 11-38 展示了如何在文本上创建反射效果。

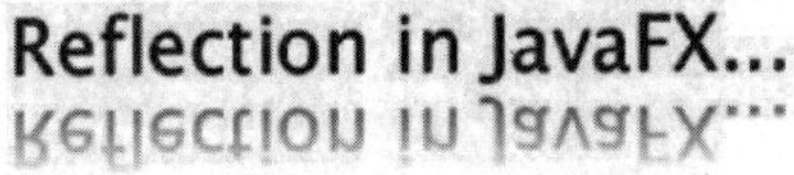

Reflection in JavaFX...

图 11.12 应用于文本的反射

Example 11-38　Text With Reflection

```
import javafx.scene.text.*;
import javafx.scene.paint.*;
import javafx.scene.effect.*;
public class HelloEffects extends Application {
    Stage stage;
    Scene scene;
    @Override public void start(Stage stage) {
        stage.show();
        scene = new Scene(new Group( ), 840, 680);
```

第 11 章 基于 JavaFX 开发动画与视觉效果

```
        ObservableList<Node> content = ((Group) scene.getRoot()).
getChildren();
        content.add(reflection());
        stage.setScene(scene);
    }
    static Node reflection() {
        Text t = new Text();
        t.setX(10.0f);
        t.setY(50.0f);
        t.setCache(true);
        t.setText("Reflection in JavaFX...");
        t.setFill(Color.RED);
        t.setFont(Font.font("null", FontWeight.BOLD, 30));
        Reflection r = new Reflection();
        r.setFraction(0.9);
        t.setEffect(r);
        t.setTranslateY(400);
        return t;
    public static void main(String[] args) {
        Application.launch(args);
    }
}
```

11.3.4 照明效果

照明效果用来模拟光源照射在给定内容上的效果,可为平面对象提供更逼真的三维外观。图 11.13 所示为文本的照明效果。Example 11-39 展示了如何在文本上创建照明效果。

JavaFX Lighting!

图 11.13 照明效果示例

Example 11-39 Text with Applied Lighting Effect

```java
import javafx.application.Application;
import javafx.collections.ObservableList;
import javafx.geometry.VPos;
import javafx.scene.effect.Light.Distant;
import javafx.scene.*;
import javafx.stage.*;
import javafx.scene.shape.*;
import javafx.scene.effect.*;
import javafx.scene.paint.*;
import javafx.scene.text.*;
public class HelloEffects extends Application {
    Stage stage;
    Scene scene;
    @Override public void start(Stage stage) {
        stage.show();
        scene = new Scene(new Group());
        ObservableList<Node> content = ((Group)scene.getRoot()).getChildren();
        content.add(lighting());
        stage.setScene(scene);
    }
    static Node lighting() {
        Distant light = new Distant();
        light.setAzimuth(-135.0f);
        Lighting l = new Lighting();
        l.setLight(light);
        l.setSurfaceScale(5.0f);
        Text t = new Text();
        t.setText("JavaFX"+"\n"+"Lighting!");
        t.setFill(Color.RED);
        t.setFont(Font.font("null", FontWeight.BOLD, 70));
        t.setX(10.0f);
        t.setY(10.0f);
        t.setTextOrigin(VPos.TOP);
        t.setEffect(l);
        t.setTranslateX(0);
        t.setTranslateY(320);
```

```
        return t;
    }
    public static void main(String[ ] args) {
        Application.launch(args);
    }
}
```

11.3.5 透视效果

透视效果用来创建二维对象的三维效果，图 11.14 所示为透视效果。

图 11.14 透视效果示例

透视效果可以将任何正方形映射到另一个正方形中，同时保持直线的直线度。与仿射变换不同，源中行的并行性不一定会保留在输出中。

> 注意：透视效果不会调整输入事件及其坐标在节点上测量包容度的任何方法；鼠标点击事件如果使用透视效果，则包含方法未定义应用于节点。

Example 11-40 是示例应用程序中的一段代码片段，展示了如何创建透视效果。

Example 11-40 Perspective Effect

```
static Node perspective( ) {
    Group g = new Group( );
    PerspectiveTransform pt = new PerspectiveTransform();
    pt.setUlx(10.0f);
    pt.setUly(10.0f);
    pt.setUrx(210.0f);
    pt.setUry(40.0f);
    pt.setLrx(210.0f);
    pt.setLry(60.0f);
    pt.setLlx(10.0f);
    pt.setLly(90.0f);
    g.setEffect(pt);
    g.setCache(true);
    Rectangle r = new Rectangle( );
    r.setX(10.0f);
    r.setY(10.0f);
```

```
            r.setWidth(280.0f);
            r.setHeight(80.0f);
            r.setFill(Color.DARKBLUE);
            Text t = new Text();
            t.setX(20.0f);
            t.setY(65.0f);
            t.setText("Perspective");
            t.setFill(Color.RED);
            t.setFont(Font.font("null", FontWeight.BOLD, 36));
            g.getChildren().add(r);
            g.getChildren().add(t);
            return g;
        }
```

图 11.15 所示为哪些坐标会影响生成的图像。

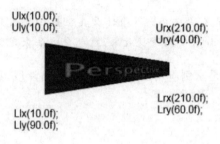

图 11.15　哪些坐标会影响生成的图像

11.3.6　创建一个效应链

有些效果有一个输入属性,可以用来创建一系列效果。效应链可以是树状结构,这是因为有些效应有两个输入,而有些效应则没有任何输入。

在图 11.16 中,反射效果被用作阴影效果的输入,这意味着矩形首先被反射

图 11.16　效应链效果示例

第 11 章 基于 JavaFX 开发动画与视觉效果

效果反射,然后阴影效果被应用到结果中。Example 11-41 依次应用了阴影和反射效果。

Example 11-41

```
import javafx.application.Application;
import javafx.collections.ObservableList;
import javafx.scene.*;
import javafx.stage.*;
import javafx.scene.shape.*;
import javafx.scene.effect.*;
import javafx.scene.paint.*;
import javafx.scene.text.*;
public class HelloEffects extends Application {
    Stage stage;
    Scene scene;
    @Override public void start(Stage stage) {
        stage.show();
        scene = new Scene(new Group());
        ObservableList<Node> content = ((Group) scene.getRoot()).getChildren();
        content.add(chainEffects());
        stage.setScene(scene);
    }
    static Node chainEffects() {
        Rectangle rect = new Rectangle();
        rect.setFill(Color.RED);
        rect.setWidth(200);
        rect.setHeight(100);
        rect.setX(20.0f);
        rect.setY(20.0f);
        DropShadow ds = new DropShadow();
        ds.setOffsetY(5.0);
        ds.setOffsetX(5.0);
        ds.setColor(Color.GRAY);
        Reflection reflection = new Reflection();
        ds.setInput(reflection);
        rect.setEffect(ds);
        return rect;
```

```
    }
    public static void main(String[ ] args) {
        Application.launch(args);
    }
}
```

　　如果将 chainEffects（ ）方法中的最后两行更改为"reflection.setInput(ds);"和"rect.rect.setEffect(reflec- tion);"，则首先将阴影效果应用于矩形，然后通过反射效果实现反射。

　　上述讨论的各种效果综合在如下 JavaFX 应用项目中。运行这个项目，将得到如图 11.17 所示的运行结果。

图 11.17　视觉效果综合示例的运行结果

转换、动画和视觉效果教程的源代码

Tutorial	Source Code
Transformations	Xylophone.java
Overview	
Animation Basics	
Tree Animation	
Example	
Creating Visual Effects	

Application Files
- NetBeans Projects
- visual_effects.zip

11.4 小结

本章介绍了如何基于 JavaFX 开发具有变换、时间轴动画以及视觉效果的 JavaFX 应用,基于示例介绍了相关的基本概念与实现原理。希望读者认真阅读每个示例的源代码,每个示例的开发原理与过程是最值得学习与借鉴的地方。

附录1　图形教程的源代码

（1）本文档中的演示应用程序及其关联的源代码文件。

Canvas Samples

Chapter	JavaFX File	NetBeans Files
Canvas	—	BasicOpsTest.zip
		CanvasDoodleTest.zip
		CanvasTest.zip
		LayerTest.zip

（2）3D MoleculeSampleApp Code。

本附录列出了用于构建 3D 示例应用程序的 MoleculeSampleApp 应用程序的源代码文件。

① Xform.java

② buildMolecule()

③ handleMouse()

④ handleKeyboard()

附录 2　WebViewSample 应用的源代码文件

- WebViewSample.java
- BrowserToolbar.css
- help.html
- Image Resources

WebViewSample.java

```java
package webviewsample;
import javafx.application.Application;
import javafx.application.Platform;
import javafx.beans.value.ObservableValue;
import javafx.collections.ListChangeListener.Change;
import javafx.concurrent.Worker.State;
import javafx.event.ActionEvent;
import javafx.event.Event;
import javafx.geometry.HPos;
import javafx.geometry.Pos;
import javafx.geometry.VPos;
import javafx.print.PrinterJob;
import javafx.scene.Node;
import javafx.scene.Scene;
import javafx.scene.control.Button;
import javafx.scene.control.ComboBox;
import javafx.scene.control.ContextMenu;
import javafx.scene.control.Hyperlink;
import javafx.scene.control.MenuItem;
import javafx.scene.image.Image;
import javafx.scene.image.ImageView;
import javafx.scene.input.MouseButton;
import javafx.scene.input.MouseEvent;
import javafx.scene.layout.HBox;
import javafx.scene.layout.Priority;
import javafx.scene.layout.Region;
import javafx.scene.paint.Color;
import javafx.scene.web.PopupFeatures;
import javafx.scene.web.WebEngine;
import javafx.scene.web.WebHistory;
```

```java
import javafx.scene.web.WebHistory.Entry;
import javafx.scene.web.WebView;
import javafx.stage.Stage;
import netscape.javascript.JSObject;
public class WebViewSample extends Application {
    private Scene scene;
    @Override
    public void start(Stage stage) {
        //创建场景
        stage.setTitle("Web View Sample");
        scene = new Scene(new Browser(stage), 900, 600, Color.web("#666970"));
        stage.setScene(scene);
        //应用CSS
         scene.getStylesheets().add("webviewsample/BrowserToolbar.css");
        //显示舞台
        stage.show();
    }
    public static void main(String[] args) {
        launch(args);
    }
}
class Browser extends Region {
    private final HBox toolBar;
    final private static String[] imageFiles = new String[] {
        "product.png", "blog.png", "documentation.png", "partners.png", "help.png"
    };
    final private static String[] captions = new String[] {
        "Products", "Blogs", "Documentation", "Partners", "Help"
    };
    final private static String[] urls = new String[] {
        "http://www.oracle.com/products/index.html", "http://blogs.oracle.com/", "http://docs.oracle.com/javase/index.html",
        "http://www.oracle.com/partners/index.html",
        WebViewSample.class.getResource("help.html").toExternalForm()
    };
    final ImageView selectedImage = new ImageView();
    final Hyperlink[] hpls = new Hyperlink[captions.length];
```

附录2　WebViewSample 应用的源代码文件

```java
    final Image[] images = new Image[imageFiles.length];
    final WebView browser = new WebView();
    final WebEngine webEngine = browser.getEngine();
    final Button toggleHelpTopics = new Button("Toggle Help Topics");
    final WebView smallView = new WebView();
    final ComboBox comboBox = new ComboBox();
    private boolean needDocumentationButton = false;
    public Browser(final Stage stage) {
      //应用 styles
      getStyleClass().add("browser");
      for (int i = 0; i < captions.length; i++) {
        //创建超文本链接
        Hyperlink hpl = hpls[i] = new Hyperlink(captions[i]);
        Image image = images[i] = new Image(getClass().
getResourceAsStream(imageFiles[i]));
        hpl.setGraphic(new ImageView(image));
        final String url = urls[i];
        final boolean addButton = (hpl.getText().equals("Help"));
        //处理事件
        hpl.setOnAction((ActionEvent e) -> {
          needDocumentationButton = addButton;
          webEngine.load(url);
        });
      }
    comboBox.setPrefWidth(60);
    //创建工具栏
    toolBar = new HBox();
    toolBar.setAlignment(Pos.CENTER);
    toolBar.getStyleClass().add("browser-toolbar");
    toolBar.getChildren().add(comboBox);
    toolBar.getChildren().addAll(hpls);
    toolBar.getChildren().add(createSpacer());
    //设置按钮的动作事件处理
    toggleHelpTopics.setOnAction((ActionEvent t) -> {
        webEngine.executeScript("toggle_visibility('help_topics')");
    });
    smallView.setPrefSize(120, 80);
    //处理 popup 窗口
    webEngine.setCreatePopupHandler(
      (PopupFeatures config) -> {
```

```java
      smallView.setFontScale(0.8);
      if (!toolBar.getChildren().contains(smallView)) {
        toolBar.getChildren().add(smallView);
      }
      return smallView.getEngine();
    });
    //处理history
    final WebHistory history = webEngine.getHistory();
    history.getEntries().addListener((Change<? extends Entry> c) -> {
      c.next();
      c.getRemoved().stream().forEach((e) -> {
        comboBox.getItems().remove(e.getUrl());
      });
      c.getAddedSubList().stream().forEach((e) -> {
        comboBox.getItems().add(e.getUrl());
      });
    });
    //设置历史记录组合框的事件
    comboBox.setOnAction((Event ev) -> {
      int offset = comboBox.getSelectionModel().getSelectedIndex()
- history.getCurrentIndex();
      history.go(offset);
    });
    //处理加载页
    webEngine. getLoadWorker ( ). stateProperty ( ). addListener
((ObservableValue<? extends State> ov, State oldState,
State newState) -> {
      toolBar.getChildren().remove(toggleHelpTopics);
      if (newState == State.SUCCEEDED) {
        JSObject win = (JSObject) webEngine.executeScript("window");
        win.setMember("app", new JavaApp());
        if (needDocumentationButton) {
          toolBar.getChildren().add(toggleHelpTopics);
        }
      }
    });
    //添加上下文菜单
    final ContextMenu cm = new ContextMenu();
    MenuItem cmItem1 = new MenuItem("Print");
    cm.getItems().add(cmItem1);
```

```
toolBar.addEventHandler(MouseEvent.MOUSE_CLICKED, (MouseEvent e)
-> {
    if (e.getButton() == MouseButton.SECONDARY) {
      cm.show(toolBar, e.getScreenX(), e.getScreenY());
    }
});
//处理打印作业
cmItem1.setOnAction((ActionEvent e) -> {
    PrinterJob job = PrinterJob.createPrinterJob();
    if (job != null) {
    webEngine.print(job);
    job.endJob();
  }
});
//加载 HomePage
webEngine.load("http://www.oracle.com/products/index.html");
//add components
getChildren().add(toolBar);
getChildren().add(browser);
}
//JavaScript 接口对象
public class JavaApp {
  public void exit() {
  Platform.exit();
  }
}
private Node createSpacer() {
  Region spacer = new Region();
  HBox.setHgrow(spacer, Priority.ALWAYS);
  return spacer;
}
@Override
protected void layoutChildren() {
  double w = getWidth();
  double h = getHeight();
  double tbHeight = toolBar.prefHeight(w);
   layoutInArea(browser, 0, 0, w, h - tbHeight, 0, HPos.CENTER, VPos.
CENTER);
   layoutInArea(toolBar, 0, h - tbHeight, w, tbHeight, 0, HPos.CENTER,
VPos.CENTER);
```

```
    }
    @Override
    protected double computePrefWidth(double height) {
      return 900;
    }
    @Override
      protected double computePrefHeight(double width) {
        return 600;
      }
    }
```

BrowserToolbar.css

```
.browser {
  -fx-background-color: #666970;
}
.browser-toolbar .hyperlink, .browser-toolbar .button, .browser-toolbar{
  -fx-text-fill: white;
}
.browser-toolbar{
  -fx-base: #505359;
  -fx-background: #505359;
  -fx-shadow-highlight-color: transparent;
  -fx-spacing: 5;
  -fx-padding: 4 4 4 4;
}
```

help.html

```
<html lang="en"><head><!-- Visibility toggle script --> <script type="text/javascript">
<!--
  function toggle_visibility(id) {
    var e = document.getElementById(id);
    if (e.style.display == 'block')
      e.style.display = 'none';
    else
      e.style.display = 'block';
  }
```

```
//-->
</script></head>
<body>
<h1>Online Help</h1>
<p class="boxtitle"><a href="#" onclick="toggle_visibility('help
_topics');"
class="boxtitle">[+] Show/Hide Help Topics</a></p>
<ul id="help_topics" style='display:none;'>
<li>Products - Extensive overview of Oracle hardware and software
products, and summary of Oracle consulting, support, and educational
services.</li>
<li>Blogs - Oracle blogging community.</li>
<li>Documentation - Landing page to start learning Java. The page
contains links to the Java tutorials, developer guides, and API
documentation.</li>
< li > Partners - Oracle partner solutions and programs. Popular
resources and membership opportunities.</li>
</ul>
< p > < a href =" about: blank" onclick =" app. exit ( )" > Exit the
Application</a></p>
</body></html>
```

Image Resources

The following images can be used to build the toolbar icons in the WebViewSample application.

Table A-1

File Name	Image
product.png	
blog.png	
documentation.png	
partners.png	
help.png	

附录3 示例源代码

SimpleSwingBrowser.java

```java
package simpleswingbrowser;
import javafx.application.Platform;
import javafx.beans.value.ChangeListener;
import javafx.beans.value.ObservableValue;
import javafx.embed.swing.JFXPanel;
import javafx.event.EventHandler;
import javafx.scene.Scene;
import javafx.scene.web.WebEngine;
import javafx.scene.web.WebEvent;
import javafx.scene.web.WebView;
import javax.swing.*;
import java.awt.*;
import java.awt.event.*;
import java.net.MalformedURLException;
import java.net.URL;
import static javafx.concurrent.Worker.State.FAILED;
public class SimpleSwingBrowser extends JFrame {
    private final JFXPanel jfxPanel = new JFXPanel();
    private WebEngine engine;
    private final JPanel panel = new JPanel(new BorderLayout());
    private final JLabel lblStatus = new JLabel();
    private final JButton btnGo = new JButton("Go");
    private final JTextField txtURL = new JTextField();
    private final JProgressBar progressBar = new JProgressBar();
    public SimpleSwingBrowser() {
        super();
        initComponents();
    }
    private void initComponents() {
        createScene();
        ActionListener al = new ActionListener() {
          @Override
          public void actionPerformed(ActionEvent e) {
            loadURL(txtURL.getText());
```

```java
            }
        };
        btnGo.addActionListener(al);
        txtURL.addActionListener(al);
        progressBar.setPreferredSize(new Dimension(150, 18));
        progressBar.setStringPainted(true);
        JPanel topBar = new JPanel(new BorderLayout(5, 0));
        topBar.setBorder(BorderFactory.createEmptyBorder(3, 5, 3, 5));
        topBar.add(txtURL, BorderLayout.CENTER);
        topBar.add(btnGo, BorderLayout.EAST);
        JPanel statusBar = new JPanel(new BorderLayout(5, 0));
        statusBar.setBorder(BorderFactory.createEmptyBorder(3, 5, 3, 5));
        statusBar.add(lblStatus, BorderLayout.CENTER);
        statusBar.add(progressBar, BorderLayout.EAST);
        panel.add(topBar, BorderLayout.NORTH);
        panel.add(jfxPanel, BorderLayout.CENTER);
        panel.add(statusBar, BorderLayout.SOUTH);
        getContentPane().add(panel);
        setPreferredSize(new Dimension(1024, 600));
        setDefaultCloseOperation(JFrame.EXIT_ON_CLOSE);
        pack();
    }
    private void createScene() {
        Platform.runLater(new Runnable() {
        @Override
        public void run() {
          WebView view = new WebView();
          engine = view.getEngine();
          engine.titleProperty().addListener(new ChangeListener<String>() {
            @Override
              public void changed(ObservableValue<? extends String> observable, String oldValue, final String newValue) {
                SwingUtilities.invokeLater(new Runnable() {
                    @Override
                    public void run() {
                       SimpleSwingBrowser.this.setTitle(newValue);
                    }
```

```java
      });
    }
  });
  engine.setOnStatusChanged(new EventHandler<WebEvent<String>
>() {
    @Override
    public void handle(final WebEvent<String> event) {
      SwingUtilities.invokeLater(new Runnable() {
        @Override
        public void run() {
          lblStatus.setText(event.getData());
        }
      });
    }
  });
   engine.locationProperty().addListener(new ChangeListener<
String>() {
    @Override
     public void changed(ObservableValue<? extends String> ov,
String oldValue, final String newValue) {
      SwingUtilities.invokeLater(new Runnable() {
        @Override
        public void run() {
          txtURL.setText(newValue);
        }
      });
    }
  });
    engine.getLoadWorker().workDoneProperty().addListener(new
ChangeListener<Number>() {
    @Override
      public void changed(ObservableValue<? extends Number>
observableValue, Number oldValue, final Number newValue) {
        SwingUtilities.invokeLater(new Runnable() {
         @Override
          public void run() {
            progressBar.setValue(newValue.intValue());
          }
        });
```

```java
            }
        });
        engine.getLoadWorker().exceptionProperty().addListener(new
ChangeListener<Throwable>() {
            @Override
            public void changed(ObservableValue<? extends Throwable>
o, Throwable old, final Throwable value) {
                if (engine.getLoadWorker().getState() == FAILED) {
                    SwingUtilities.invokeLater(new Runnable() {
                        @Override
                        public void run() {
                            JOptionPane.showMessageDialog(panel, (value !=
null) ? engine.getLocation() + "\n" + value.getMessage(): engine.
getLocation() + " \ nUnexpected error.", " Loading error...",
JOptionPane.ERROR_MESSAGE);
                        }
                    });
                }
            }
        });
        jfxPanel.setScene(new Scene(view));
    }
});
}
public void loadURL(final String url) {
    Platform.runLater(new Runnable() {
        @Override
        public void run() {
            String tmp = toURL(url);
            if (tmp == null) {
                tmp = toURL("http://" + url);
            }
            engine.load(tmp);
        }
    });
}
private static String toURL(String str) {
    try {
        return new URL(str).toExternalForm();
```

```java
            } catch (MalformedURLException exception) {
                return null;
            }
        }
        public static void main(String[ ] args) {
            SwingUtilities.invokeLater(new Runnable() {
            @Override
            public void run() {
                SimpleSwingBrowser browser = new SimpleSwingBrowser();
                browser.setVisible(true);
                browser.loadURL("http://tup.com.cn");
                }
            });
        }
}
```

Converter.java

```java
package converter;
import javafx.application.Application;
import javafx.beans.property.DoubleProperty;
import javafx.beans.property.SimpleDoubleProperty;
import javafx.collections.FXCollections;
import javafx.collections.ObservableList;
import javafx.scene.Scene;
import javafx.scene.layout.VBox;
import javafx.stage.Stage;
public class Converter extends Application {
    /**
     * @param args the command line arguments
     */
    public static void main(String[] args) {
        launch(args);
    }
    private final ObservableList<Unit> metricDistances;
    private final ObservableList<Unit> usaDistances;
    private final DoubleProperty meters = new SimpleDoubleProperty(1);
    public Converter() {
        //Create Unit objects for metric distances, and then
```

```java
        //instantiate a ConversionPanel with these Units.
        metricDistances = FXCollections.observableArrayList(new Unit("Centimeters", 0.01), new Unit("Meters", 1.0), new Unit("Kilometers", 1000.0));
        //Create Unit objects for U.S. distances, and then
        //instantiate a ConversionPanel with these Units.
        usaDistances = FXCollections.observableArrayList(new Unit("Inches", 0.0254), new Unit("Feet", 0.305), new Unit("Yards", 0.914), new Unit("Miles", 1613.0));
    }
    @Override
    public void start(Stage stage) {
        VBox vbox = new VBox(new ConversionPanel("Metric System", metricDistances, meters), new ConversionPanel("U.S. System", usaDistances, meters));
        Scene scene = new Scene(vbox);
        stage.setTitle("Converter");
        stage.setScene(scene);
        stage.show();
    }
}
```

ConversionPanel.java

```java
package converter;
import java.text.NumberFormat;
import java.text.ParseException;
import javafx.beans.InvalidationListener;
import javafx.beans.property.DoubleProperty;
import javafx.collections.ObservableList;
import javafx.scene.control.ComboBox;
import javafx.scene.control.Slider;
import javafx.scene.control.TextField;
import javafx.scene.control.TitledPane;
import javafx.scene.layout.HBox;
import javafx.scene.layout.VBox;
import javafx.util.StringConverter;
public class ConversionPanel extends TitledPane {
    final static int MAX = 10000;
```

```java
        private ComboBox<Unit> comboBox;
        private Slider slider;
        private TextField textField;
        private DoubleProperty meters;
        private NumberFormat numberFormat;
        private InvalidationListener fromMeters = t -> {
          if (!textField.isFocused()) {
            textField.setText(numberFormat.format(meters.get() /
            getMultiplier()));
          }
        };
        private InvalidationListener toMeters = t -> {
          if (!textField.isFocused()) {
            return;
          }
          try {
            meters. set ( numberFormat. parse ( textField. getText ( )).
doubleValue() * getMultiplier());
          }
          catch (ParseException | Error | RuntimeException ignored) {
          }
        };
        public ConversionPanel (String title, ObservableList< Unit > units,
DoubleProperty meters) {
          setText(title);
          setCollapsible(false);
          numberFormat = NumberFormat.getNumberInstance();
          numberFormat.setMaximumFractionDigits(2);
          textField = new TextField();
          slider = new Slider(0, MAX, 0);
          comboBox = new ComboBox(units);
          comboBox.setConverter(new StringConverter<Unit>() {
              @Override
              public String toString(Unit t) {
                  return t.description;
              }
              @Override
              public Unit fromString(String string) {
```

```java
            throw new UnsupportedOperationException("Not supported yet.");
        }
    });
    VBox vbox = new VBox(textField, slider);
    HBox hbox = new HBox(vbox, comboBox);
    setContent(hbox);
    this.meters = meters;
    comboBox.getSelectionModel().select(0);
    meters.addListener(fromMeters);
    comboBox.valueProperty().addListener(fromMeters);
    textField.textProperty().addListener(toMeters);
    fromMeters.invalidated(null);
    slider.valueProperty().bindBidirectional(meters);
}
/**
 * 返回当前选定测量单位的乘数(单位/米)
 * @return
 */
public double getMultiplier() {
    return comboBox.getValue().multiplier;
}
}
```

参 考 文 献

[1] 宋波. Java 应用开发教程[M]. 北京：电子工业出版社，2002.
[2] 宋波，董晓梅. Java 应用设计[M]. 北京：人民邮电出版社，2002.
[3] 宋波. Java Web 应用与开发教程[M]. 北京：清华大学出版社，2006.
[4] 宋波，刘杰，杜庆东. UML 面向对象技术与实践[M]. 北京：科学出版社，2006.
[5] 埃克尔. Java 编程思想[M]. 4 版. 陈昊鹏，译. 北京：机械工业出版社，2007.
[6] 刘斌，费冬冬，丁薇. NetBeans 权威指南[M]. 北京：电子工业出版社，2008.
[7] 宋波. Java 程序设计——基于 JDK 6 和 NetBeans 实现[M]. 北京：清华大学出版社，2011.
[8] Raoul-Gabriel Urma，等. Java 8 实战[M]. 北京：人民邮电出版社，2016.
[9] 千锋教育高教产品研发部. Java 语言程序设计[M]. 2 版. 北京：清华大学出版社，2017.
[10] 赫伯特·希尔德特. Java 9 编程参考官方大全[M]. 10 版. 北京：清华大学出版社，2018.
[11] 林信良. Java 学习笔记[M]. 北京：清华大学出版社，2018.
[12] 关东升. Java 编程指南[M]. 北京：清华大学出版社，2019.
[13] 凯·S.霍斯特曼. Java 核心技术 卷Ⅰ基础知识(原书第 11 版)[M]. 北京：机械工业出版社，2019.
[14] Kishori Sharan. Learn JavaFX 8：Building User Experience and Interfaces with Java 8[M]. Apress，2015.
[15] 宋波. Java 程序设计——基于 JDK 9 和 NetBeans 实现[M]. 北京：清华大学出版社，2022.